Zinc and Human Health

Zinc and Human Health

**Results of Recent Trials
and Implications for
Program Interventions and Research**

Edited by Kenneth H. Brown and Sara E. Wuehler

L'Initiative
micronutriments

The Micronutrient
Initiative

Published by the Micronutrient Initiative
PO Box 8500, Ottawa, ON, Canada K1G 3H9
http://www.micronutrient.org/

© International Development Research Centre 2000

Canadian Cataloguing in Publication Data

Main entry under title :
Zinc and human health : results of recent trials and implications for program interventions and research

Includes bibliographical references.
Co-published by the International Development Research Centre.
ISBN 1-894217-13-6

1. Zinc in the body — Congresses.
2. Zinc — Physiological effect — Congresses.
3. Zinc — Metabolism — Congresses.
I. Brown, Kenneth H.
II. Wuehler, Sara E.
III. Micronutrient Initiative (Association)
IV. International Development Research Centre (Canada)

QP535.Z6Z56 2000   612'.01524   C00-980399-8

IDRC Books endeavours to produce environmentally friendly publications. All paper used is recycled as
well as recyclable. All inks and coatings are vegetable-based products.

Published in association with the
International Development Research Centre
PO Box 8500, Ottawa, ON, Canada
http://www.idrc.ca

Canadä

Table of Contents

Table of Contents

Preface

A conference entitled "Zinc and human health: results of recent intervention trials and implications for programmatic interventions and program-linked research" was convened at the University of California, Davis, 21–23 October 1999, under the sponsorship of the university's Program in International Nutrition, in collaboration with The Micronutrient Initiative. The objectives of the conference were to review the results of recent zinc intervention trials, focusing on the impact of enhanced zinc status on pregnancy outcome, children's growth and development, and risk of morbidity from infections; to summarize the implications of these studies with regard to appropriate program interventions, with emphasis on dietary diversification or modification, supplementation, and fortification; and to identify critical gaps in current knowledge that might impede implementation of program interventions.

The conference had two major components. The first half of the meeting was devoted to individual presentations by experts in the field of zinc nutrition, who summarized current knowledge on zinc metabolism, assessment of zinc status, estimates of the global prevalence of zinc deficiency, complications of zinc deficiency, and the range of program approaches available to enhance zinc status. The second half of the conference was based on small-group discussions, the ultimate objective of which was to prepare guidelines for the development of zinc intervention programs and to identify particular issues that require further research to facilitate program implementation.

The organization of this report corresponds to that of the conference. The first part of the report contains background information on zinc nutrition and its relationship to human health, as well as estimates of the likely global prevalence of zinc deficiency. The detailed scientific review papers presented during this portion of the conference will be published separately but are summarized here. The second section of the report presents the recommendations of the discussion groups. We hope that the background information in this report will prove useful and that the report itself will stimulate policymakers to initiate public health programs designed to control the serious problem of zinc deficiency in high-risk populations.

Kenneth H. Brown   France Bégin
Sara E. Wuehler   Mahshid Lotfi
University of California, Davis   The Micronutrient Initiative

May 2000

Acknowledgments

The meeting stemmed from discussions between Venkatesh Mannar of The Micronutrient Initiative (MI) and Kenneth H. Brown of University of California, Davis on the need to increase awareness of the global situation regarding zinc deficiency, the currently available interventions to counteract this deficiency, and their program implications. The meeting was planned with the help of Kenneth Brown, Bo Lönnerdal, and other faculty members of the University of California, Davis, and Mahshid Lotfi and France Bégin of The Micronutrient Initiative. The present meeting report, which summarizes the presentations and working group discussions, was prepared by Kenneth Brown and Sara Wuehler of the University of California, Davis and was reviewed by the conference participants and by scientific staff of MI, whose comments were incorporated. D'Ann Finley provided editorial support.

Administrative support was provided by Betty Alce and Debbie Montgomery (MI) and by Diane Vandepeute (University of California, Davis).

Executive Summary

A conference entitled "Zinc and human health: results of recent intervention trials and implications for programmatic interventions and program-linked research" was convened at the University of California, Davis, 21–23 October 1999, under the sponsorship of the university's Program in International Nutrition, in collaboration with The Micronutrient Initiative, with three primary goals: to review the results of recent zinc intervention trials, focusing on the impact of enhanced zinc status on child survival and functional performance; to summarize the implications of these studies with regard to appropriate program interventions, with emphasis on dietary modification, supplementation, and fortification; and to identify critical gaps in current knowledge that might impede implementation of program interventions.

The conference participants concluded that zinc deficiency is a common public health problem, with nearly half the world's population at risk of inadequate zinc intake. People who depend primarily on plant-based diets with either low zinc content or poor zinc bioavailability, such as residents of countries in South and Southeast Asia, sub-Saharan Africa, northern Africa, and the eastern Mediterranean, as well as of some countries in Latin America, have a particularly high risk of inadequate dietary intake and consequent deficiency of zinc. The critical functional consequences of poor zinc status are adverse outcomes of pregnancy, retardation of growth in childhood, impairment of immune function, and, secondarily, increased rates of morbidity and possibly mortality from infectious disease and abnormal neurobehavioural development. Because of the widespread occurrence of zinc deficiency and its severe complications, there was strong consensus among the conference participants that action should be taken at the earliest possible opportunity to control zinc deficiency in countries where it is likely to be a problem. High-risk population groups include pregnant and lactating women and young children, who have relatively high zinc requirements.

Strategies for the alleviation of zinc deficiency include dietary diversification or modification (or both), supplementation, and fortification, along with general health programs to reduce infections that contribute to poor absorption or excessive losses of zinc. When possible, policymakers should identify opportunities to integrate zinc interventions into ongoing primary health care and other existing nutrition and public health programs rather than attempting to establish new, independent programs to control zinc deficiency. To stimulate this process, efforts should be undertaken as soon as possible to assist countries in assessing the zinc status of their populations. One practical approach that was recommended for population assessment is the review or collection of representative data on dietary intake of absorbable zinc by high-risk segments of the population. Information on the amounts of total and absorbable zinc in the food supply can also provide indirect evidence about the risks associated with inadequate zinc intake. Because of the lack of suitable biomarkers of zinc status, it is perhaps more difficult to assess the actual zinc status of a population and to monitor the impact of intervention programs than it is for other nutrients. However, available information suggests that the change in mean serum zinc concentration of the target population can provide useful evidence that an intervention has affected the population's zinc status.

Additional research is needed to support the development and implementation of zinc intervention programs. Selected research priorities include development of simple, rapid, reliable, and low-cost techniques to assess the zinc status of individuals and populations; evaluation of the effects of domiciliary and commercial food-processing techniques on zinc absorption and zinc status; examination of how different chemical and physical forms and amounts of zinc supplements and fortificants affect their acceptability, absorption, net impact on zinc status, and risk of any adverse effects on copper and iron status; and evaluation of the effectiveness and cost–benefit analyses of the foregoing intervention strategies. More detailed recommendations regarding program implementation and priorities for program-linked research are provided in the report itself.

Zinc in the Environment and in Biology

Zinc is a bluish white metallic element (atomic number 30, atomic weight 65.4), which makes up about 0.02% of the earth's crust and is the 23rd most abundant element. Because of its nature as a transitional element in the periodic table, zinc possesses certain chemical properties that make it especially useful and important in biological systems. Specifically, zinc is able to constitute strong yet readily exchangeable and flexible complexes with organic molecules, which enables it to modify the three-dimensional structure of nucleic acids, specific proteins, and cellular membranes and to influence the catalytic properties of many enzyme systems and intracellular signaling. Zinc is associated with more than 50 distinct metallo-enzymes, which have a diverse range of functions, including the synthesis of nucleic acids and specific proteins, such as hormones and their receptors (Cousins 1996). For these reasons, zinc plays a central role in cellular growth, differentiation, and metabolism. Of further interest is zinc's absence of redox properties, which allows it to be transported in biological systems without inducing oxidant damage, as can occur with other trace elements, such as iron and copper.

Prasad (1990) reviewed the early history of research into the biological importance of zinc. To summarize from his report, zinc was first recognized as an essential nutrient for microorganisms more than 125 years ago. Appreciation of its essentiality for higher plants, rats and mice, poultry, and swine followed, during the period from the 1920s through the 1950s. Despite these observations, researchers remained skeptical about the possibility of zinc deficiency in humans because of the element's ubiquity in the environment. Nevertheless, evidence of human zinc deficiency began to emerge in the 1960s, when cases of zinc-responsive dwarfism and delayed sexual maturation were first reported in Egyptian adolescents. Since then, clinical studies of children with acrodermatitis enteropathica, an inborn error of zinc metabolism resulting in poor zinc absorption and consequent severe zinc deficiency, have confirmed the critical role of zinc in physical growth and in gastrointestinal and immune function (Moynahan 1974). Moreover, as described below, zinc intervention trials have produced positive growth responses and reduced rates of infection in high-risk children in vulnerable populations.

Zinc Metabolism

Zinc is absorbed through the small intestine, which also regulates whole-body zinc homeostasis through changes in both the fractional absorption of dietary zinc and the excretion of endogenous zinc in pancreatic juice and other gastrointestinal secretions (Jackson 1989). Some zinc is also lost from the body through urine, menstrual flow, semen, and sloughed skin, nails, and hair, although the quantity lost through these other routes is small relative to that lost through gastrointestinal excretion. As with intestinal excretion, the urinary elimination of zinc can be affected by zinc status (Baer and King 1984), although this effect is less consistent and may occur only with more severe or prolonged dietary restriction. Fecal zinc excretion is also increased during diarrhea (Castillo-Duran et al. 1988), which may contribute to zinc deficiency in areas with high rates of enteric infections.

The total-body zinc content of adult humans ranges from about 1.5 to 2.5 g, most of which is intracellular, primarily in muscle, bone, liver, and other organs (Jackson 1989). Approximately 90% of the body's zinc reserves turn over slowly and are therefore not readily available for metabolism. The remaining zinc constitutes the so-called rapidly exchangeable pool of zinc, which is thought to be particularly important for maintaining the zinc-dependent functions of human biological systems. The rapidly exchangeable zinc can move into and out of the plasma compartment within a period of about 3 days. The size of this pool is sensitive to the amounts of zinc absorbed from the diet, and a reasonably constant dietary supply is thought to be necessary to satisfy the normal requirements of zinc for maintenance and growth.

Less than 0.2% of total-body zinc content circulates in the plasma, which has a mean concentration of approximately 15 µmol/L (about 100 µg/dL). In the plasma, zinc

3

is bound to albumin and, to a lesser extent, α_2-macroglobulin and oligopeptides (Cousins 1996). Because the concentration of zinc in tissues, such as muscle and liver, is approximately 50 times greater than in plasma, small differences in uptake or release of zinc from these peripheral sites can have a profound effect on the plasma zinc concentration. For these reasons, plasma zinc concentrations do not reliably indicate total-body zinc stores under all circumstances. For example, release of zinc from muscle tissue that is catabolized during starvation can result in transient, seemingly paradoxical, elevations in plasma zinc (Henry and Elmes 1975). In contrast, consumption of standard meals, or glucose alone, induces a postprandial reduction in plasma zinc concentration, even though dietary zinc intake and tissue reserves may be adequate (Hambidge et al. 1989). Other factors that influence plasma zinc concentration are hypoalbuminemia, which affects absorption and transport of zinc (Smith et al. 1978); intestinal diseases that interfere with zinc absorption (Cousins 1989); pregnancy (Swanson and King 1983); infection (Beisel et al. 1973; Falchuk 1977; Brown 1998); and other forms of stress, such as tissue injury imposed by surgery (Shenkin 1995) and strenuous physical exercise (Lukaski et al. 1984).

Zinc and Human Function

Since the early reports of human zinc deficiency, and particularly during the past 10–15 years, a considerable number of well-designed clinical trials have examined the relationships between zinc supplementation and human health. Zinc is especially important during periods of rapid growth, both pre- and postnatally, and for tissues with rapid cellular differentiation and turnover, such as the immune system and the gastrointestinal tract.

Critical functions affected by zinc nutriture include pregnancy outcome, susceptibility to infection, physical growth, and neurobehavioural development, among others.

Pregnancy Outcome

Reproductive functions that have been examined in relation to zinc status are duration of pregnancy (incorporating data for spontaneous abortion); fetal growth; the timing, sequence, and efficiency of labour and delivery; and the incidence of stillbirths and congenital malformations. This voluminous literature has been reviewed by several authors, including Hurley and Baly (1982), Apgar (1985), Keen and Hurley (1989), Tamura and Goldenberg (1996), Caulfield et al. (1998), and Caulfield et al. (1999). Although there are clear relationships between zinc nutriture and each of these outcomes in animal models (Hurley and Baly 1982; Keen and Hurley 1989), especially when zinc deficiency is severe, the results of human studies have been less consistent, possibly because of small sample sizes, other inadequacies in study design, and the difficulty in accurately classifying an individual's zinc status. In some zinc intervention trials, the subjects were unselected women who were unlikely to be deficient in zinc (Mahomed et al. 1989; Jønsson et al. 1996), so the negative results of these trials are not surprising.

In almost all of the human trials, supplementation began no earlier than the second trimester of pregnancy, so there is very little information from humans on the effect on pregnancy outcome of zinc nutriture during early pregnancy.

A small clinical trial was carried out in 56 British women who were thought to be at risk of delivering infants with intrauterine growth retardation because of low body weight, smoking, Asian origin, or previous delivery of a baby with this condition (Simmer et al. 1991). Zinc-supplemented women had significantly lower rates of induced labour (13% vs 50%), cesarean section (7% vs 32%), and infants with intrauterine growth retardation (7% vs 27%), although the small sample size and differential dropout rates weaken the inferences that can be drawn from this study. A second, frequently cited controlled intervention trial was completed in 585 low-income African-American women with plasma zinc concentrations below the population median (Goldenberg et al. 1995). Overall, there was a significant increase (by 126 g) in birth weight with zinc supplementation, and among women with initial body mass index below the study median (i.e., $< 26 \, kg/m^2$), birth weight increased by 248 g. In contrast, in a recently completed trial in 1016 Peruvian women at risk of zinc deficiency, zinc supplementation had no impact on the duration of gestation or on infant birth weight (Caulfield et al. 1999). Importantly, however, differences in fetal and newborn behaviour were detected. Specifically, fetuses of zinc-supplemented mothers were more active in utero at 36 weeks of gestation (Merialdi et al. 1998). These differences were interpreted as indicating a positive effect of maternal zinc supplementation on fetal neurobehavioural development. Other infant outcomes, such as postnatal behavioural development and susceptibility to infection, are still under analysis.

In summary, there is considerable information from studies of animals with severe zinc deficiency and suggestive evidence from human intervention trials to indicate that maternal zinc nutriture can influence several aspects of reproductive function and pregnancy outcome. Additional studies are needed of humans in different settings and under various study conditions to define more precisely which women are likely to benefit from interventions to enhance their zinc status. Because of evidence from animal studies indicating that even a few days of low zinc intake at different stages of pregnancy can affect pregnancy outcome, human studies are needed in which supplemental zinc is provided throughout pregnancy, including the first trimester. The impact on pregnancy outcome of maternal zinc status at the time of conception also needs to be addressed.

Morbidity

During the past few years a number of studies have been completed to determine the effects of zinc supplementation on the incidence, and in some cases the severity and duration, of diarrhea, pneumonia, and malaria. To examine the consistency of results and to increase the statistical power of results obtained in individual studies, the results of the trials examining diarrhea and pneumonia outcomes have been pooled recently in several secondary analyses. Seven community-based trials of continuous zinc supplementation were used in a pooled analysis of the preventive effects of zinc (Bhutta, Black et al. 1999). In all seven of the studies, the incidence and prevalence of diarrhea were reduced with zinc supplementation, which resulted in a pooled odds ratio (OR) of 0.82 (95% confidence interval [CI] 0.72–0.93) for the effect of zinc on diarrheal incidence and a pooled OR of 0.75 (95% CI 0.63–0.88) for the effect of zinc on diarrheal prevalence. Four studies also provided information on incidence of pneumonia, yielding a pooled OR of 0.59 (95% CI 0.41–0.83) for the effect of zinc. Thus, continuous zinc supplementation resulted in substantial and consistent reductions in both diarrhea and acute infections of the lower respiratory tract.

In a separate set of pooled analyses, the effect of supplementary zinc supplied as an adjunct to other therapy was evaluated in children with acute or persistent diarrhea (Zinc Investigators' Collaborative Group 2000). The pooled results of three trials of children with acute diarrhea indicated that those who received zinc supplements recovered 15% more rapidly (95% CI 6% to 22%) than those in the control groups. In an additional three studies of patients with persistent diarrhea, children who received supplements recovered 29% sooner (95% CI 6% to 57%) than those in control groups. In the latter studies there was also a 42% reduction (95% CI 10% to 63%) in treatment failure or death among children who were given supplemental zinc. Thus, therapeutic use of zinc for adjunctive treatment of children with either acute or persistent diarrhea reduced the duration of illness and the risk of treatment failure. Because of the known negative association between diarrheal prevalence and growth velocity, zinc therapy might also reduce the impact of these illnesses on growth.

Research has now been completed in which supplemental zinc was provided for just 2 weeks beginning at the time of initiation of treatment for either acute or persistent diarrhea in Bangladeshi children (Roy et al. 1999). Growth and risk of infection were monitored for 2 months after the presenting episode of illness. Notably, zinc-treated children continued to have reduced prevalence of diarrhea for 1–2 months after the supplements were discontinued. Thus, there may also be some benefit of intermittent zinc supplementation in high-risk populations.

Physical Growth

A total of 25 controlled clinical trials of the effect of zinc supplementation on children's growth were summarized in a recently completed meta-analysis (Brown et al. 1998). Overall, zinc supplementation had a small, but highly significant, impact on children's height increments, with an average effect size of 0.22 standard deviation (SD) units. This effect was present for the subgroup of studies with mean initial height-for-age Z-scores less than –2.0, but not for those with mean initial height-for-age Z-scores of –2.0 or greater. The height response to zinc supplementation was unrelated to the dosage schedule or to the duration of supplementation. In studies with initially stunted children, the effect size of zinc supplementation was moderately large, averaging 0.49 SD units. Zinc supplementation also had a small, but highly significant, impact on children's weight increments, with an average effect size of 0.26 SD units. Interestingly, the effect of zinc supplementation on change in weight was negatively associated with mean initial plasma zinc levels. In the subset of studies with low mean initial plasma zinc concentrations (<80 µg/dL [<12 µmol/L]), the impact of zinc supplementation was moderately large. The effect of zinc supplementation on children's growth may be due to its direct impact on hormonal mediators of growth (Ninh et al. 1996) or its effects on appetite (O'Dell and Reeves 1989) or risk of infection (Bhutta, Black et al. 1999). Notably, in none of these studies was there significantly reduced growth velocity or any other indication of complications associated with zinc supplementation. On the basis of this extensive research experience, there seems to be strong evidence of a relationship between zinc deficiency and growth stunting. In settings with high rates of stunting, underweight, or low plasma zinc concentrations, or any combination of these, programs to enhance zinc status may be useful interventions to increase children's growth and to decrease current rates of nutritional stunting.

Neurobehavioural Development

As recently reviewed by Black (1998), results of multiple studies in experimental animals indicate that a broad range of neurobehavioural abnormalities can occur with zinc deficiency. For example, studies in fetuses of zinc-deprived pregnant rats demonstrated neuronal degeneration and a reduction in brain size (Prohaska et al. 1974; McKenzie et al. 1975), and studies in prepubertal monkeys with moderate zinc restriction found lower spontaneous motor activity and reduced performance of tasks that required visual attention (Golub et al. 1985; Golub et al. 1994; Golub et al. 1996).

Information available from studies of zinc deficiency or zinc supplementation in humans is less consistent, perhaps in part because of the broad range of subjects observed and research designs employed. Nevertheless, a number of zinc-related behavioural abnormalities have also been described in humans. For example, Aggett (1989) noted that changes in mood, loss of affect, and emotional lability were pronounced enough to serve as warning signs of early zinc deficiency in patients receiving total parenteral nutrition or children with acrodermatitis enteropathica. Henkin et al. (1975) induced severe acute zinc deficiency in six adults by administering histidine and observed the following progression of symptoms: anorexia, dysfunction of smell and taste, irritability, depression, and anger. These subjects also displayed lethargy, sleepiness, and decreased ability to perform simple mental arithmetic and to interpret proverbs.

Prospective studies comparing the behaviour of zinc-supplemented infants and children with that of control subjects have produced variable results. Friel et al. (1993) found higher motor development scores among zinc-supplemented very-low-birth-weight Canadian infants than controls. By contrast, subsequent studies of low-birth-weight Brazilian infants (Ashworth et al. 1998) and older infants and toddlers in Guatemala (Bentley et al. 1997) and India (Sazawal et al. 1996) found no differences in motor development, although greater activity and responsiveness were observed with zinc supplementation. Merialdi et al. (1998) described greater activity in the fetuses of zinc-supplemented Peruvian mothers. Studies of zinc supplementation of young school-age children in Guatemala (Cavan et al. 1993) and Canada (Gibson et al. 1989) found no differences in attention span after supplementation. However, Chinese school children who received supplemental zinc and other micronutrients performed better on a battery of neuropsychologic tests than those who received only the other nutrients (Penland et al. 1997; Sandstead et al. 1998).

Researchers generally agree that the mechanisms by which zinc deprivation induces behavioural changes are unclear and that further study is needed to determine at what stage zinc deficiency influences behaviour, whether these effects are reversible, how long zinc supplementation is needed to prevent the risks, and whether other limiting nutrients must be provided for zinc supplementation to be effective.

Dietary Sources and Bioavailability of Zinc

In addition to the previously described effect of an individual's zinc status on zinc absorption, the total zinc content of the diet and the bioavailability of zinc from the diet's food components also influence the efficiency of zinc absorption. Discounting the effect of zinc status, zinc absorption is determined largely by its solubility in the intestinal lumen, which in turn is affected by the chemical form of zinc and the presence of specific inhibitors and enhancers of absorption. The major inhibitor of zinc absorption is myoinositol hexaphosphate (phytate), which is present in many plant foods, especially cereals and legumes, and irreversibly binds zinc under conditions present in the intestinal lumen. Organs and flesh of mammals, fowl, fish, and crustaceans are the richest food sources of zinc, and these foods do not contain phytate; therefore, they are particularly good sources of absorbable zinc (Table 1). Eggs and dairy products are also free of phytate, although they have a slightly lower zinc content than organs and flesh foods. Most cereals and legumes have an intermediate level of zinc, but their high phytate content reduces the amount of zinc available for absorption. When these staples are fermented (as occurs with leavened breads and with porridges prepared from fermented cereals), the fermenting organisms produce phytases that break down the phytates, which increases the amount of absorbable zinc. Rice and starchy roots and tubers have lower zinc contents than legumes and cereals other than rice. For the most part, fruits and vegetables are not rich sources of zinc, although some green leafy vegetables, such as spinach, have a fairly high zinc density, albeit of uncertain bioavailability.

**Table I. Zinc and phytate contents of different foods, and
estimated amount of zinc available for absorption.***

Food type	Zinc content		Phytate content		Estimated absorbable zinc* (mg/100 g)
	In mg/ 100 g	In mg/ 100 kcal	In mg/ 100 g	Phytate–zinc molar ratio	
Liver, kidney (of beef or poultry)	4.2–6.1	2.7–3.8	0	0	2.1–3.1
Meat (beef, pork)	2.9–4.7	1.1–2.8	0	0	1.4–2.4
Poultry (e.g., chicken, duck)	1.8–3.0	0.6–1.4	0	0	0.9–1.5
Seafood[†]	0.5–5.2	0.3–1.7	0	0	0.2–2.6
Eggs (chicken, duck)	1.1–1.4	0.7–0.8	0	0	0.6–0.7
Dairy products (cow milk, cheese)	0.4–3.1	0.3–1.0	0	0	0.2–1.6
Seeds, nuts (e.g., sesame, pumpkin, almond)	2.9–7.8	0.5–1.4	1760–4710	22–88	0.3–0.8
Bread (made with white flour, yeast)	0.9	0.3	30	3	0.4
Whole-grain cereal (e.g., wheat, maize, brown rice)	0.5–3.2	0.4–0.9	211–618	22–53	0.1–0.3
Beans, lentils (e.g., soy, kidney bean, chickpea)	1.0–2.0	0.9–1.2	110–617	19–56	0.1–0.2
Refined cereal grains (e.g., white flour, white rice)	0.4–0.8	0.2–0.4	30–439	16–54	0.1
Fermented cassava root	0.7	0.2	70	10	0.2
Tubers	0.3–0.5	0.2–0.5	93–131	26–31	<0.1–0.2
Vegetables	0.1–0.8	0.3–3.5	0–116	0–42	<0.1–0.4
Fruits	0–0.2	0–0.6	0–63	0–31	<0.1–0.2

*Amount of zinc available for absorption estimated as 45% to 55% if phytate–zinc molar ratio < 5, as 30% to 35% if phytate–zinc molar ratio = 5–15, and as 10% to 15% if phytate–zinc molar ratio > 15.
†Excluding oysters.

To estimate the likely absorption of zinc from a mixed diet, a phytate–zinc molar ratio can be calculated as follows: (phytate content of foods/660)/(zinc content of foods/65.4), where 660 and 65.4 represent the molecular or atomic weight of phytate and zinc, respectively. Information on the phytate content of various foods has been compiled in several databases (Oberleas and Harland 1981; Oberleas 1983; Reddy et al. 1989). It is generally considered that diets with a phytate–zinc molar ratio greater than 15 have relatively poor zinc biovailability, those with a phytate–zinc molar ratio between 5 and 15 have medium zinc biovailability, and those with a phytate–zinc molar ratio less than 5 have relatively good zinc biovailability (WHO 1996). The relationship between zinc intake, the phytate–zinc molar ratio of the diet, and zinc absorption has been described by WHO (1996), which defined three categories of diets, with high, medium, or low zinc availability, based on the proportion of energy from animal sources, the types of processing of cereals, the amounts of inorganic calcium salts, and the phytate–zinc molar ratio. WHO (1996) estimated that about 45% to 55% of the amount of dietary zinc is absorbed from a high-bioavailability diet, 30% to 35% from a medium-bioavailability diet, and 10% to 15% from a low-bioavailability diet (Fig. 1), depending on the zinc content of the meal (and assuming a typical intake of approximately 3–5 mg zinc per meal). Because of remaining uncertainties in the estimates of zinc absorption from different diets by individuals with different zinc statuses, further work is needed to validate the ability of dietary data to predict zinc status.

On the basis of her own and others' research, Sandström has recently summarized the results of previous studies of the fractional and net absorption of zinc under

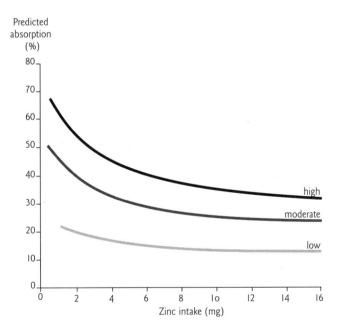

Fig. I. Predicted fractional absorption of dietary zinc as a function of zinc intake and according to level of zinc bioavailability in the diet. Data from WHO (1996).

different dietary conditions (Sandström and Ceberblad 1980, 1987; Sandström et al. 1980; Farah et al. 1984; Lönnerdal et al. 1984; Valberg et al. 1984; Flanagan et al. 1985; Nävert et al. 1985; Sandström et al. 1985; Sandström et al. 1987; Sandström et al. 1989). The results of this analysis are shown in Table 2 and are summarized in the following paragraphs. When zinc sulfate was consumed in aqueous solution by fasting individuals, the mean fractional absorption ranged from 46% to 74%. Although fractional absorption decreased with higher intakes of zinc, the net absorption of zinc was greater when the total consumption of zinc was increased. When the same zinc solutions were consumed with small amounts of rice crisps or wheat bran, fractional zinc absorption declined considerably, to about 5% to 40% of the original levels (resulting in final fractional absorptions of 2% to 22%). The fractional absorption of zinc was also reduced when other minerals were included in the aqueous supplement, although the decrement in fractional absorption of zinc with these additional minerals was less dramatic than with the phytate-containing foods.

Fractional absorption of zinc from milk formulas (about 32%) was substantially greater than from soy formulas (about 12%). The absorption coefficients did not change much when more zinc was added to the formulas, so fortification resulted in greater net absorption of zinc. Zinc absorption from white and whole-meal breads varied with phytate content, and fractional absorption in both types of bread decreased to less than half the original levels when they were fortified. Notably, the net absorption of zinc increased threefold when white bread was enriched with 3.1 mg zinc chloride to produce a total meal content of 3.5 mg zinc, but not when sufficient zinc was added to the same amount of (phytate-containing) whole-meal bread to produce a similar final zinc content. When meals containing animal products (chicken meat) were enriched with 3.3 mg zinc per meal, fractional absorption fell by only about one-fourth and net absorption increased by 30% to 50%. Similar results were observed with turkey meat. Addition of 55 g of chicken containing 0.7 mg zinc to 70 g of bean meal containing 2.7 mg zinc increased both the fractional and the net absorption of zinc over absorption levels with the beans alone. Incubation of dough with yeast for 16 h before baking reduced the phytate–zinc molar ratio of bread containing wheat bran from 17 to 4 and doubled the absorption of zinc.

Table 2. Zinc absorption from aqueous solutions, zinc supplements, and enriched foods or meals.*

Form of zinc	Zinc content (mg)	Phytate–zinc molar ratio	Zinc absorption (mean ± SD) %	mg	Reference
Aqueous solution					
Taken in the fasting state					
Zinc sulfate	2.6	0	74±5	1.9	Sandström et al. 1985
Zinc sulfate	2.6	0	73±0	1.9	Sandström et al. 1987
Zinc sulfate	13	0	46±13	6	Sandström et al. 1987
Zinc sulfate	15	0	58±13	8.8	Farah et al. 1984
Taken with meal					
Zinc sulfate + rice crisps	15 + 0.2	<1	22±7	3.3	Farah et al. 1984
Zinc sulfate + wheat bran	15 + 2	3	2±1	0.4	Farah et al. 1984
Multielement supplement					
Taken in the fasting state					
Zinc citrate	15	0	38±7	5.7	Sandström and Cederblad 1987
Taken with meal					
Zinc citrate + low-phytate meal	15 + 1.6	<1	10±2	1.7	Sandström and Cederblad 1987
Enriched meals					
Cow's milk formula	1.2	0	32±1	0.4	Lönnerdal et al. 1984
Cow's milk formula + zinc, 6 mg/L	3.9	0	33±2	0.9	Lönnerdal et al. 1984
Soy formula	1.7	6	14±1	0.2	Lönnerdal et al. 1984
Soy formula + zinc, 8 mg/L	5.3	4	11±1	0.4	Lönnerdal et al. 1984
White bread	0.4	3	38	0.2	Sandström et al. 1980
White bread + zinc chloride	0.4 + 3.1	<1	13	0.5	Sandström et al. 1980
Whole-meal bread	1.3	23	17	0.2	Sandström et al. 1980
Whole-meal bread + zinc chloride	1.3 + 2.2	15	8	0.3	Sandström et al. 1980
Chicken meat (low phytate)	1.3	5	36±7	0.5	Sandström and Cederblad 1980
Chicken meat + zinc chloride	1.3 + 3.3	2	24±5	1.1	Sandström and Cederblad 1980
Turkey	2	0	39±8	0.8	Flanagan et al. 1985
Turkey	3	0	29±6	0.9	Flanagan et al. 1985
Turkey + zinc chloride	2 + 1	0	28±8	1.1	Valberg et al. 1984
Beef meal	4.6	2	20±5	0.9	Sandström and Cederblad 1980
Beef meal + zinc chloride	4.6 + 3.3	<1	15±3	1.2	Sandström and Cederblad 1980
Food fermentation					
Unleavened bran bread	1.9	17	10±4	0.2	Nävert et al. 1985
Bran bread leavened for 16 h	1.8	4	20±5	0.4	Nävert et al. 1985
Food enrichment					
Bean meal (70 g beans dry weight)	2.7	15	22±6	0.6	Sandström et al. 1989
Bean meal (70 g beans dry weight) + 55 g chicken	3.4	12	32±5	1.1	Sandström et al. 1989

SD = standard deviation.
*Information compiled by B. Sandström.

In summary, zinc was absorbed most efficiently from aqueous solutions and from meals containing animal products. Absorption was considerably less from phytate-containing meals. Fortification of foods with exogenous zinc generally produced a small reduction in fractional absorption, but a positive impact on net absorption. However, fortification of foods with a high phytate–zinc molar ratio had only a small effect on net zinc absorption.

Zinc Requirements

Estimates of zinc requirements have been developed by several expert committees, including those convened by WHO (1996), the National Research Council (1989) of the USA, and the Panel on Dietary Reference Values of the Committee on Medical Aspects of Food Policy (1991) of the UK. The WHO committee estimated the physiological zinc requirements of adults as the sum of the amounts needed for tissue growth, maintenance, metabolism, and replacement of endogenous losses. Because intestinal and urinary losses change in relation to zinc status, two separate estimates of these requirements were developed by WHO (1996), as follows. The so-called "basal requirements" of 1.0 mg/d in adult males and 0.7 mg/d in adult females refer to the amounts needed to balance the aforementioned physiological requirements of individuals who are fully adapted to low zinc intake. Because this level of intake leaves no reserve for adaptation to any further decrease in intake, the second estimate was developed to provide a greater margin of safety. This estimate, which is referred to as the "normative physiological requirement," accounts for the fact that zinc absorption must be about 40% greater in individuals who are not yet adapted to low intakes, in order to balance fecal and urinary losses. Hence, the normative zinc requirement was set at 1.4 mg/d for adult males and at 1.0 mg/d for adult females.

Once the normative zinc requirement is applied to a particular category of individuals, the dietary requirement must be adjusted to account for the estimated percentage of zinc absorption from the diet, as follows: dietary requirement = (normative requirement)/(percent absorption from usual diet). The WHO committee (WHO 1996) further suggested that population dietary recommendations be established at a level such that individuals who consume 2 SD units less than the population average would be able to satisfy their normative requirement. This would assure that in populations with a mean intake at the recommended population level nearly all individuals would be able to satisfy their zinc needs. The WHO committee assumed a coefficient of variation in dietary intake of 25% and therefore proposed a population mean "normative dietary requirement" at a level such that 50% (or 2 SD) less than this mean would be equivalent to the age- and sex-specific normative requirement. For example, if the normative requirement for adult males is 1.4 mg/day and the diet is of medium bioavailability (i.e., there is approximately 30% zinc absorption), then the normative dietary requirement is 1.4/0.3 or 4.7 mg. The population mean dietary requirement is

Table 3. Recommended population dietary zinc intake to meet normative physiological requirements (WHO 1996).

Population group and age range* (years)	Recommended dietary zinc intake (mg)		
	High bioavailability†	Medium bioavailability†	Low bioavailability†
Infants			
0–0.5	na	na	na
0.5–1	3.3	5.6	11.1
Children			
1–3	3.3	5.5	11.0
3–6	3.9	6.5	12.9
6–10	4.5	7.5	15.0
Men			
10–12	5.6	9.3	18.7
12–15	7.3	12.1	24.3
15–18	7.8	13.1	26.2
18–60+	5.6	9.4	18.7
Women			
10–12	5.0	8.4	16.8
12–15	6.1	10.3	20.6
15–18	6.2	10.2	20.6
18–60+	4.0	6.5	13.1
Pregnant women, any age			
(mean for all trimesters)	6.0	10.0	20.0
Lactating women, any age			
0–5 months after birth	7.3	12.2	24.3
≥6 months after birth	5.8	9.6	19.2

na = not applicable, because breast-fed infants < 6 months of age are assumed to be able to satisfy zinc requirements from human milk only.

*For each age range, subjects up to but not including the age indicated by the upper value were included. For example, the range 1–3 years includes children from exactly 1 year of age to 3 years less 1 day (i.e., does not include children exactly 3 years old).

†Low bioavailability: 45% to 55% absorption of zinc intake (phytate–zinc molar ratio < 5); medium bioavailability: 30% to 35% absorption of zinc intake (phytate–zinc molar ratio = 5–15); low bioavailability: 10% to 15% absorption of zinc intake (phytate–zinc molar ratio > 15).

therefore set at 9.4 mg/d to assure that individuals who consume only 50% of the population mean dietary requirement would satisfy their normative dietary requirement.

In addition to the foregoing reasoning applied to adults, the estimates of zinc requirements for children take into consideration the amounts needed for accretion of newly formed tissue and the proportionately greater endogenous losses of zinc per unit body weight in children. To account for the latter phenomenon, the endogenous losses per unit of basal metabolic rate in adults (rather than per unit body weight) were applied to children. The population mean dietary requirements suggested by WHO (1996) are presented by age, sex, and physiological status in Table 3. The WHO recommendations assume that a fully breast-fed infant less than 6 months of age is able to meet his or her zinc needs from breast milk alone.

The US National Research Council (1989) similarly assumed that the zinc requirements of adults could be established by measuring endogenous losses and adjusting for the amount of zinc absorbed from the diet. However, the US recommendations considered only one level of zinc absorption from a mixed diet (20% of intake). These recommendations are generally consistent with the WHO recommendations for a diet with low to medium zinc bioavailability. The UK recommendations (Panel on Dietary Reference Values of the Committee on Medical Aspects of Food Policy 1991) applied a similar conceptual framework but assumed a single figure of 30% absorption from the diet. Hence, the estimated UK reference zinc intake is generally less than the US estimate and is consistent with the WHO estimate for a medium-bioavailability diet.

Assessment of Zinc Status

Several authors have reviewed the range of techniques that have been proposed for evaluating the zinc nutriture of individuals and populations (King 1990; Aggett and Favier 1993; Hambidge and Krebs 1995). Most researchers agree that efficient regulation of zinc homeostasis complicates the diagnosis of zinc deficiency and excess and that there are no generally accepted, reliable biomarkers of individual zinc status. Only when zinc deficiency is relatively severe is it possible to detect changes in tissue zinc concentrations. For these reasons, the definitive diagnosis of zinc deficiency is usually based on a high index of suspicion (as motivated by low dietary intake, poor bioavailability of zinc, or suggestive clinical signs [such as growth retardation, delayed sexual maturation, dermatitis, behavioural changes, and defects in immune function], or some combination of these) and documentation of a functional response to zinc supplementation. Obviously, the need to conduct a therapeutic trial to identify a functional response to supplementation complicates the ability to diagnose zinc deficiency and increases the cost of this assessment, especially because several months may be needed to measure certain functional responses, such as physical growth, with confidence.

Of the alternative techniques that have been proposed for direct evaluation of a population's zinc status, the ones that seem most promising for field application are assessment of the adequacy of zinc intake and measurement of the population's mean plasma zinc concentration. A manual describing a dietary method developed for determining the adequacy of zinc (and iron) intakes in developing countries has been compiled by Gibson and Ferguson (1999). Dietary assessment requires quantitative measurement of food intake from a representative sample of the population of interest, knowledge of the zinc content of these foods, and appraisal of the likely absorption of zinc from the mixed diets (Murphy et al. 1992; Gibson and Ferguson 1998a).

Dietary intake can be assessed quantitatively by a number of different methods, the simplest of which for population assessment is a modified interactive 24-h dietary recall (Gibson 1990; Gibson and Ferguson 1998a, 1999). Once the daily food intake is known, the total zinc intake can be estimated by multiplying the amount of each of the foods that are consumed by its zinc content, as recorded in local food composition tables or in databases available internationally, such as the US Department of Agriculture (USDA) food composition database (USDA 1999). For developing countries the World Food Dietary Assessment Program can be used (Bunch and Murphy 1994). The nutrient database associated with this system contains food composition values for 1800 foods from six countries (Egypt, Kenya, Mexico, Senegal, India, and Indonesia) and provides data for 53 nutrients and antinutritional factors, including zinc, iron, dietary fibre, and phytate. Local databases are theoretically advantageous because the zinc content of foods can vary according to soil conditions, agronomic practices, and local food-processing techniques, although in many cases local food composition tables contain fewer food items and fewer replicate analyses per item. Moreover, these local tables frequently omit analyses of zinc and phytate. To estimate the amount of zinc available for absorption from the diet, the diet can be categorized according to its phytate–zinc molar ratio, as described above.

Use of the mean plasma zinc concentration to assess a population's zinc status requires proper collection and processing of blood from a representative sample of the population and analysis of the zinc concentration in the plasma or serum. Despite the difficulties in interpreting the plasma zinc concentration of individual subjects, several pieces of evidence suggest that the mean plasma zinc concentration of a group of individuals may provide useful information on the zinc status of the population from which that sample is derived. For example, when the zinc intake of groups of volunteer study subjects is severely restricted, the plasma zinc concentrations diminish within a fairly short period (Baer and King 1984). Furthermore, results from a recent meta-analysis of zinc intervention trials indicate that the mean plasma zinc concentration of subjects in individual studies predicted the magnitude of the response in weight gain following zinc supplementation (Brown et al. 1998). When the initial mean plasma zinc concentration was greater than about 80 μg/dL (about 12 μmol/L), there was no response to zinc supplementation. With decreasing plasma zinc concentrations, there was, in general, a progressively greater response to zinc supplementation.

Moreover, almost all studies found a significant increase in the plasma zinc concentrations following supplementation, suggesting that this indicator could also be used to assess successful delivery of zinc supplements.

Zinc Toxicity

Zinc has low toxicity, although acute symptoms of nausea, vomiting, diarrhea, fever, and lethargy may be observed when large doses (i.e., ≥ 1 g) are consumed. When zinc intake exceeds physiological needs by reasonably small amounts, homeostasis can be maintained by increased endogenous fecal and urinary excretion. However, if excessive zinc intake continues for prolonged periods, absorption of other trace elements, especially copper, can be impaired. For example, intake of supplements providing 50 mg zinc per day for 6 weeks produced changes in erythrocyte copper–zinc superoxide dismutase, an indicator of copper status (Fischer et al. 1984; Yadrick et al. 1989). At higher doses of zinc (160–660 mg/d), anemia and changes in immune function and lipoprotein metabolism have been observed, in addition to abnormal indices of copper status (Porter et al. 1977; Hooper et al. 1980; Chandra 1984; Patterson et al. 1985). Further studies are needed to specify the level of zinc intake at which any undesirable effects on copper metabolism, hematologic indices, immune function, and lipoprotein metabolism begin to occur. Moreover, it needs to be determined whether the recommended upper limit of zinc intake should be modified when this nutrient is consumed with other foods of various phytate contents.

Estimates of the Global Prevalence of Zinc Deficiency

Several authors have presented cogent arguments on the likelihood of widespread zinc deficiency in low-income countries (Sandstead 1991; Shrimpton 1993; Gibson 1994). However, quantitative estimates of the percentage of the global population at risk of inadequate zinc nutriture and specific information on the prevalence of deficiency in particular settings are still lacking, in large part because of the aforementioned difficulties in assessing individual zinc status. This lack of information has been a major limiting factor in attempting to convince policymakers of the need to develop programs to reduce the rate of zinc deficiency.

One indirect method that can be used for estimating global rates of zinc deficiency is analysis of previously collected information on the total daily per capita amount of zinc in the national food supply in relation to the population's theoretical zinc requirements. Data on the amounts of major food commodities available for human consumption are listed for most countries in the Food and Agriculture Organization's (FAO) food balance sheets (FAO 1998). Despite the inherent weaknesses in this type of national-level statistics, the food balance sheets do provide reasonably reliable information on the total amounts and types of foods available in a broad range of countries. By calculating the amount of zinc present in these foods and the amounts that are potentially absorbable, as estimated from their phytate–zinc ratios, it is possible to assess whether the food supply is adequate to satisfy the population's theoretical requirements for zinc.

Zinc in the Global Food Supply

To estimate the zinc content of the global food supply, information was downloaded from the FAO's food balance sheets (FAO 1998). The food balance sheets provide information on the annual amounts of 95 major food commodities that are available for human consumption in 178 countries. The amount of zinc potentially available for absorption from these foods was also estimated, by using WHO guidelines based on the phytate–zinc ratio of the food supply (WHO 1996). Finally, the mean daily per capita amount of zinc in the food supply was compared with the population's mean daily per capita normative dietary zinc requirement, weighted according to the age and sex distribution of the population (UN 1994).

Table 4 displays regional data on the mean daily per capita availability of the following items in the food supply of 178 countries: total energy, total zinc, zinc density, phytate, phytate–zinc molar ratio, estimated absorbable zinc, and zinc as a percentage of the weighted mean daily per capita zinc requirement. The regions are ranked in descending order by the amount of absorbable zinc in the food supply. The mean daily per capita amount of zinc in the national food supply ranges from about 11–12 mg in the more affluent countries of Western Europe, North America, and the western Pacific to about 7–9 mg in the poorer regions of South and Southeast Asia, northern Africa and the eastern Mediterranean, and sub-Saharan Africa. The zinc content of the national food supplies in China and Latin America is intermediate. The total zinc content of national food supplies is strongly associated with total energy content and percentage of energy provided by animal sources (data not shown). Notably, in the wealthier countries more than half the zinc is provided by animal sources, whereas only 15% to 25% of zinc comes from these sources in the poorer countries, which leads to sizeable differences in the phytate–zinc molar ratios among regions and approximately

Table 4. Mean (± standard deviation) daily per capita amounts of energy, zinc, and phytate in the food supply of 178 countries, by region.*

Region	No. of countries	Population (millions)	Energy (kcal/d)	Zinc (mg/d)	Zinc density (mg/1000 kcal)	Phytate (mg/d)	Phytate–zinc molar ratio	Absorbable zinc (mg/d)	% requirement†	Estimated % population at risk of low zinc intake
Western Europe	20	457	3410±135	12.4±1.3	3.6±0.4	1596±391	13.2±4.8	3.2±1.2	137±46	8.0±17.1
USA and Canada	2	305	3546±164	12.2±0.5	3.5±0.1	1542±58	12.5±0.1	2.9±0.1	122±5	0.9±0.2
Eastern Europe	27	413	2971±255	10.8±1.3	3.6±0.3	1567±211	14.5±2.2	2.1±0.6	87±22	12.8±18.1
Western Pacific	13	223	2902±273	11.8±1.0	4.1±0.3	2123±444	18.1±4.7	2.0±1.2	88±46	18.6±16.3
Latin America and Caribbean	35	498	2743±313	9.9±2.0	3.6±0.4	2111±808	21.1±6.0	1.5±1.1	73±43	45.8±26.7
China‡	2	1262	2743±33	10.9±0.2	4.0±0.1	2074±36	18.8±0.6	1.5±0.2	69±7	21.4±1.5
Southeast Asia	10	504	2556±226	9.0±0.9	3.5±0.1	2248±586	24.5±4.6	1.1±0.2	56±7	71.2±14.2
Sub-Saharan Africa	46	581	2203±379	9.3±2.0	4.3±0.8	2530±645	26.9±3.7	1.0±0.2	59±13	68.0±25.9
North Africa and eastern Mediterranean	17	342	2806±450	8.7±1.6	3.1±0.6	2206±524	25.1±3.5	1.0±0.3	55±12	73.5±20.5
South Asia	6	1297	2351±99	7.6±0.6	3.2±0.2	2068±263	26.9±1.7	0.8±0.1	47±4	95.4±2.1
All regions	178	5882	2706±434	10.0±2.0	3.6±0.5	2045±504	21.3±6.0	1.5±0.9	72±34	48.9±36.8

*Analyses assume that wheat is consumed as 99% white flour (75% extraction) and 1% whole wheat: rice is consumed as white rice: other cereals as whole grains. No assumptions have been made with regard to fermentation of foods. Information on national food supplies is drawn from national food balance sheets prepared by the Food and Agriculture Organization (FAO 1998).
†Adjusted for phytate content (n = 163).
‡Includes Hong Kong.

threefold differences in the estimated amount of zinc that is available for absorption (Table 4). The food supplies of Western Europe and North America provide about 150% of the weighted mean per capita normative dietary requirement, whereas the food supplies of South and Southeast Asia, northern Africa and the eastern Mediterranean, and sub-Saharan Africa provide only about 50% to 60% of this requirement.

The percentage of the population at risk of deficiency was estimated for each country, as follows. The mean normative dietary requirement of zinc was assumed to be 50% of the WHO population dietary requirement as calculated above and was assumed to be fixed. Availability of food sources of zinc to individuals was assumed to follow a Gaussian distribution, with the standard deviation equal to 25% of the mean, as suggested by WHO (1996). The percentage of the population at risk was calculated as the area under the normal curve to the left of the normative physiological requirement. According to this approach, the global food supply places nearly half the world's population at risk of low zinc intake (i.e., the national food supplies of the countries analyzed provide nearly half the people in the world with less than the weighted mean per capita normative dietary requirement of their respective countries). The percentage of the national population at risk of low zinc intake ranges from a low of 1% to 13% in countries of Europe and North America to a high of 68% to 95% in South and Southeast Asia, Africa, and the eastern Mediterranean region.

A number of weaknesses in this approach to estimating the global prevalence of zinc deficiency must be recognized. First, the estimates of the amount of zinc in the food supply are only as good as the information provided in the national food balance sheets. Because the amount of zinc in national food supplies is strongly correlated with the total amount of food energy, any underestimates in the available food supply will result in overestimates of the number of individuals at risk of low intake and vice versa. Second, the food balance sheets provide national-level data, but no information is given on the distribution of the food supply among and within households. The coefficient of variation in population zinc intake was assumed to be 25%, as suggested by WHO (1996), but if the actual variability in intake is greater than this, the current analyses would tend to underestimate the percentage of individuals at risk of low intake in countries with food zinc supplies above the requirement and overestimate the percentage of individuals at risk of low intake in countries with food zinc supplies below the requirement. Third, the WHO estimates of age- and sex-specific normative physiological requirements assume that these are fixed, when in fact they vary among individuals. It is conceivable that some individuals satisfy their physiological needs with lower zinc intakes. Finally, other possible flaws in this approach can result from inaccuracies in the food composition database, failure to correct adequately for the effects of food processing on zinc and phytate contents, and uncertainties in the estimated physiological requirements for zinc and bioavailability of zinc from different diets. Also, it must be recognized that the total amount of food in the national food supply is almost certainly greater than the amounts actually consumed, and this difference between food availability and food consumption is probably greater in the more affluent countries.

Notwithstanding the uncertainties in the estimates of the adequacy of zinc in national food supplies, the data do provide reasonable insights into the countries and regions that are likely to be at greatest risk of low intake of absorbable zinc and consequent zinc deficiency. Notably, these countries are ones that tend to have elevated rates of infant and child mortality, low birth weight, and postnatal malnutrition (UNICEF 1999).

Programmatic Approaches to Control Zinc Deficiency

As with other micronutrients, there are several possible direct and indirect strategies that could be applied to control zinc deficiency. Direct approaches include diversification or modification of the existing diet to increase the consumption or absorption of zinc, supplementation with zinc-containing compounds, and fortification of suitable food vehicles with additional zinc. Indirect approaches include public health programs aimed at reducing conditions that may adversely affect zinc status, such as diarrhea or intestinal helminths.

Dietary Diversification or Modification

Possible ways of enhancing zinc status through changes in the diet are by promoting increased consumption of foods, such as animal products, that are particularly rich in zinc or that promote zinc absorption or by reducing intake of inhibitors of zinc absorption, generally by food-processing techniques. In both cases, successful interventions require knowledge of current dietary patterns, food-processing techniques, and other factors, such as cultural attitudes and food costs, that may influence food choices and methods of preparation. To maximize acceptability and sustainability, any suggested modifications to the diet should not increase substantially the cost of the diet or the time required to prepare it. To ensure that the recommendations will be feasible, understandable, and acceptable to consumers, it is desirable to use a participatory research approach, in which the target population is directly involved in developing the ultimate recommendations for any proposed dietary diversification or modification and the educational messages used to disseminate this information. Because of the relatively high cost of animal products, attempts to increase their consumption may need to be coupled with new household production strategies or education to promote use of cheaper animal products and preferential consumption by the most vulnerable members of the household.

In situations where the diet is composed of cereals and legumes with reasonably high zinc contents but low amounts of absorbable zinc (because of the phytate content of these foods), interventions should focus on reducing phytate intake. Enzymatic hydrolysis of phytate can be initiated by soaking, germinating, or fermenting cereals. Soaking and germination activate endogenous cereal phytases, whereas fermentation with lactic acid bacteria generates microbial phytases. In both cases, these enzymes partially break down phytate into lower inositol phosphates (i.e., inositol phosphate-1, inositol phosphate-2, and inositol phosphate-3), which are no longer capable of binding zinc (Lönnerdal et al. 1989). These processing techniques may have other advantages, such as reducing cereal viscosity in the case of germination and decreasing proliferation of bacterial pathogens in the case of lactic acid fermentation. Soaking also permits partial removal, by diffusion, of water-soluble phytate from certain cereals, such as maize, even without enzymatic hydrolysis. These processing techniques can often be combined to enhance their overall impact on reducing the phytate content of the diet (Gibson and Ferguson 1998b).

Supplementation

Supplementation is the provision of additional zinc, usually in the form of some chemical (or pharmaceutical) compound rather than as food. The major advantages of supplementation are the low cost of zinc compounds relative to the cost of the appropriate amounts of food that would have to be consumed to provide similar amounts of absorbable zinc and the ability to target selected subgroups of a population. Issues that must be considered in the development of supplementation programs include the physical and chemical forms of the zinc compound, the dose of zinc and the frequency of consumption, the inclusion of other micronutrients in the supplement, the manner of delivery (with or without foods), the packaging and distribution system, and the possible risk of toxic effects. Most of the existing experience with zinc supplementation is derived from research trials; we are not aware of any attempts to deliver zinc supplements in large-scale programs.

Zinc supplements can be formulated as liquids (generally sugar-containing, flavoured syrups), tablets, powders, or capsules. Because children less than about 5 years of age are usually unable to swallow whole tablets and capsules, zinc supplements that are given directly to youngsters must be provided either as liquid supplements or as soluble tablets or powders that can be mixed with a beverage or food. Liquid supplements are more expensive than dry ones, and simultaneous inclusion of several nutrients in liquid preparations may be limited by problems of chemical compatibility. Recently, zinc has also been included in a mixture of micronutrients provided as a high-fat spread, to be consumed alone or added to some component(s) of the existing diet (Briend et al. 1999). This latter approach falls between supplementation and fortification and is advantageous both because of the favourable palatability of the product and the lack of chemical interaction among nutrients in the lipid medium. Another "hybrid" approach is the use of single-dose sachets of dry micronutrients ("sprinkles") that are added to food at the time of serving.

Available zinc compounds vary in their water solubility at neutral pH (Table 5). Water-soluble preparations are generally better absorbed than insoluble ones, especially in individuals with low production of gastric acid (Henderson et al. 1995). This is of particular concern in areas with high rates of malnutrition or infection with *Helicobacter pylori*, both of which impair secretion of gastric acid (Gilman et al. 1988; Dale et al. 1998). Because soluble salts are more likely to have an unpleasant taste, they are better accepted when mixed with flavouring agents or delivered with food. As discussed previously, zinc is generally better absorbed from liquid solutions than from foods. If liquid supplements are given immediately before or after a meal, any inhibitors of absorption present in the meal would probably reduce zinc uptake from the supplement. Zinc absorption can also be depressed by inclusion of other nutrients in a multiple-micronutrient supplement. However, there is some evidence that the effect of iron on zinc absorption is minimal when their molar ratio is close to 1 : 1 (Sandström et al. 1985). Additional information is needed on absorption of zinc from different supplements (with and without other micronutrients) provided with or apart from typical meals in different locales. Because the major expense of a zinc supplementation program is attributable to the packaging and distribution of the supplement, inclusion of other nutrients in the zinc supplement (or inclusion of zinc in an existing

Table 5. Characteristics of zinc compounds.*

Compound	Colour	Taste	Odour	Solubility in water (at 20°C)	Other solvents
Zinc acetate	White	Astringent	Slight odour of acetic acid	Soluble	Alcohol
Zinc carbonate	White	Astringent	Odourless	Practically insoluble	Dilute acid, alkalis, solution of NH_4^+ salts
Zinc chloride[†]	White	Astringent	Odourless	Soluble	Alcohol, glycerol, ether, HCl, acetone
Zinc citrate	White	na	Odourless (in powder form)	Slightly soluble	Dilute mineral acids, alkali hydroxides
Zinc gluconate[†]	White	na	Odourless	Soluble	Alcohol
Zinc lactate	White	na	Odourless	Slightly soluble	na
Zinc methionine	White	Slightly sour and bitter	Vanilla	Soluble	na
Zinc oxide[†]	White, gray, or yellowish white	Bitter, astringent	Odourless	Practically insoluble	Dilute acids, alkalis, ammonia, ammonium carbonate
Zinc stearate[†]	White	na	Faint	Insoluble	Acids, benzene
Zinc sulfate anhydrous[†]	Colourless	na	Odourless	Soluble	na
Zinc sulfate heptahydrate[†‡]	Colourless	Astringent	Odourless	Soluble	Glycerol (insoluble in alcohol)

na = information not available.

*Adapted from Budavari et al. 1996; Davidsson 1999; P. Ranum, American Ingredients Company, Kansas City, MO, USA, personal communication.

[†]Compound generally recognized as safe (GRAS designation) (National Archives and Records Administration 1999).

[‡]Monohydrate does not clump so is more convenient than heptahydrate in warm climates.

preparation designed to control some other nutrient deficiency) can be accomplished with relatively little additional cost.

With regard to the dose of zinc, levels of 1–4 mg of zinc per day were administered to newborns in the small number of available trials, and (with one exception) 3–40 g/d (median 10 mg/d) was provided to older children (Table 6). Somewhat higher doses of zinc (usually 20–50 mg/d, median 20 mg/d) were offered for short periods (usually 2–4 weeks) during studies of severely malnourished children or children with diarrhea to compensate for assumed preexisting deficits or excessive fecal losses during illness. Zinc supplementation trials during pregnancy supplied 15–45 mg/d (median 22.5 mg/d). There have been very few studies comparing the efficacy and potential toxicity of different doses of zinc. Because of concerns regarding the adverse effect of zinc

Table 6. Doses and chemical forms of zinc used in supplementation trials.

Study	Country	Dose (mg*)	Frequency (d/week)	Chemical form	Mean age (years)
Newborns					
Walravens and Hambidge 1976	USA	4	7	Sulfate	0.01
Friel et al. 1993	Canada	1.5	7	Sulfate	0.1
Castillo-Duran et al. 1995	Chile	3	7	Acetate	0
Lira et al. 1998	Brazil	1	6	Sulfate	0
Children					
Ronaghy et al. 1974	Iran	32	6	Carbonate	13
Hambidge et al. 1978	USA	14	5	Sulfate	4.4
Hambidge et al. 1979	USA	3	7	Oxide	4.8
Walravens et al. 1983	USA	10	7	Sulfate	4.2
Smith et al. 1985	Australia	20	5	Acetate	10
Castillo-Duran et al. 1987	Chile	11.3	7	Acetate	0.6
Gatheru et al. 1988	Kenya	40	7	Sulfate	1.7
Gibson et al. 1989	Canada	10	7	Sulfate	6.4
Walravens et al. 1989	USA	5.7	7	Sulfate	1.5
Schlesinger et al. 1992	Chile	11	7	Chloride	0.6
Udomkesmalee et al. 1992	Thailand	25	5	Gluconate	9.3
Walravens et al. 1992	France	5	7	Sulfate	0.5
Bates et al. 1993	Gambia	20	2	Gluconate	1.5
Cavan et al. 1993	Guatemala	10	5	Amino acid	6.8
Nakamura et al. 1993	Japan	79	7	Sulfate	5.9
Shrivastava et al. 1993	India	5.6	7	Sulfate	1.3
Castillo-Duran et al. 1994	Chile	10	7	Sulfate	8.7
Dirren et al. 1994	Ecuador	10	6	Sulfate	2.6
Ninh et al. 1996	Viet Nam	10	7	Sulfate	1.5
Sempértegui et al. 1996	Ecuador	10	7	Sulfate	3.5
Friis et al. 1997	Zimbabwe	40	3.5	Sulfate	10.1
Rosado et al. 1997	Mexico	20	5	Methionine	2.4
Ruel et al. 1997	Guatemala	10	7	Sulfate	0.5–0.75†
Ruz et al. 1997	Chile	10	7	Sulfate	3.3
Sazawal et al. 1997, 1998	India	10	7	Gluconate	0.5–2.9†
Shankar et al. 1997	Papua New Guinea	10	7	Gluconate	0.5–2.9†
Heinig et al. 1998	USA	5	7	Sulfate	0.3
Meeks Gardner et al. 1998	Jamaica	5	7	Sulfate	1.2
Rivera et al. 1998	Guatemala	10	7	Sulfate	0.6
Sandstead et al. 1998	China	20	6	na	7.5
Penny et al. 1999	Peru	10	7	Gluconate	1.6

(continued)

Table 6 concluded.

Study	Country	Dose (mg*)	Frequency (d/wk)	Chemical form	Mean age (years)
Malnourished children‡					
Khanum et al. 1988	Bangladesh	40	7	Sulfate	2.4
Simmer et al. 1988	Bangladesh	50	7	Sulfate	3.2
Golden and Golden 1992	Jamaica	4.5	7	Acetate	1.2
Acute diarrhea§					
Sazawal et al. 1995	India	20	7	Gluconate	0.5–2.9†
Roy et al. 1999	Bangladesh	20	7	Acetate	0.9
Persistent diarrhea§					
Penny et al. 1997	Peru	20	7	Gluconate	1.6
Roy et al. 1998	Bangladesh	20	7	Acetate	0.25–2.0†
Bhutta, Nizami et al. 1999	Pakistan	3 mg/kg	7	Sulfate	0.5–3.0†
Pregnancy					
Jameson and Ursing 1976	Sweden	45	7	Sulfate	25
Hunt et al. 1984	USA	20	7	Acetate	24
Hunt et al. 1985	USA	20	7	Sulfate	16
Ross et al. 1985	South Africa	4.6–12.9	7	Gluconate	na
Kynast and Saling 1986	Germany	20	7	Aspartate	26
Cherry et al. 1989	USA	30	7	Gluconate	18
Mahomed et al. 1989	UK	20	7	Sulfate	26
Simmer et al. 1991	UK	22.5	7	Citrate	27
Garg et al. 1993	India	45	7	Sulfate	na
Goldenberg et al. 1995	USA	25	7	Sulfate	23
Jønsson et al. 1996	Denmark	44	7	na	28
Osendarp et al. 1998	Bangladesh	30	7	na	na
Caulfield et al. 1999	Peru	15	7	Sulfate	24

na = information not reported.
*Except as indicated otherwise.
†Mean not reported.
‡Children hospitalized for treatment of severe malnutrition.
§For all studies of acute and persistent diarrhea, the subjects were children.

on copper status, the lowest effective dose of zinc should be used preferentially. Additional dose–response studies are needed to determine the minimal effective dose of zinc in different settings and for different age ranges.

The frequency of administration of supplements must also be considered, although there is little information comparing different delivery schedules. In one study, 70 mg of zinc (as gluconate) was delivered twice weekly to rural Gambian preschool children (Bates et al. 1993). The zinc-treated group had greater increases in arm circumference, improvement in one indicator of intestinal mucosal permeability, and a decreased rate of self-referral to clinic for episodes of malaria, although there were no other significant effects of zinc on physical growth, immune function, or incidence of diarrhea or infection. In another study, in Viet Nam, the effects of daily or weekly administration of a multiple micronutrient supplement, containing either a 5-mg daily dose of zinc (provided 5 d/week, as sulfate) or a 17-mg weekly dose of the same zinc compound, were compared with placebo (Thu et al. 1999). There were positive effects of both forms of supplementation on serum zinc concentration and linear growth of stunted children, regardless of the frequency with which the supplement was delivered. Less frequent (i.e., weekly) distribution of a supplement would be considerably cheaper, logistically simpler, and possibly more acceptable to the consumer, so additional studies of the optimal dosing schedule would be valuable.

Fortification

Fortification is the addition of a nutrient or mixture of nutrients to a processed food, beverage, or condiment, either to restore or standardize its original composition or to augment its natural nutrient content. Particular food vehicles, such as staple foods or condiments, may be chosen for fortification because they are nearly universally consumed by individuals in a population, whereas other special-purpose foods, such as infant cereals, may be selected because they are consumed by a specific segment of the population at high risk of deficiency, which permits targeting of the intervention. In both cases, the usual amount of consumption of the food vehicle must be measured or estimated to establish the appropriate level of fortification. Studies of total dietary intake are also useful to estimate the shortfall in absorbable zinc consumption. Sensory trials are required to assure that there are no untoward effects of fortification on the taste, appearance, or shelf life of the food vehicle, and studies of nutrient bioavailability are desirable to confirm that the fortificant is adequately absorbed from the final product. Major programmatic concerns include selecting the appropriate food vehicle and fortificant, establishing the proper dose of fortificant, and monitoring the level of fortification and its impact on the population's nutritional status.

Several publications have reviewed technical issues concerning mineral fortification of foods (SUSTAIN 1998; Hurrell 1999), so these points will not be reiterated here. Zinc compounds that are available for fortification are, for the most part, the same as those described for supplementation (Table 5). As discussed in the preceding section on supplementation, water-soluble compounds, such as zinc sulfate and zinc acetate, are generally better absorbed than less soluble compounds, such as zinc oxide (Davidsson 1999). At the levels that might be considered for fortification programs, there is no evidence that zinc has adverse effects on the sensory properties of bread (Kilic et al. 1998) or other food vehicles tested (Davidsson 1999).

The proper level of zinc fortification is that which would suitably increase intake of absorbable zinc by individuals at risk of deficiency without imposing a risk of excessive intake on other members of the population who consume the fortified product. There is currently some uncertainty regarding the safe upper limit of zinc intake. WHO (1996) recommends consumption of no more than about 60 mg zinc per day. The US Environmental Protection Agency (EPA 1998) has established upper levels of safe intake for a number of minerals, including zinc. For this purpose, the EPA uses animal and human data (when available) to establish a "reference dose for chronic oral exposure," which is defined as "an estimate — with uncertainty spanning perhaps an order of magnitude — of daily exposure to the human population that is likely to be without an appreciable risk of deleterious effects during a lifetime." For a number of reasons, the theoretical upper limits that have been established by the EPA for exposure to essential minerals are extremely conservative, in many cases resulting in overlap with the recommended dietary allowances for the same nutrients. Moreover, the proposed upper limits do not take bioavailability into consideration. As Olin (1998) has stated, "The reference dose is designed to be not so much predictive of toxicity as protective of health." Thus, it is likely that the reference dose can be exceeded by reasonably small amounts without risk of toxicity, although additional data are needed to confirm this assumption. The EPA reference dose for long-term oral zinc exposure is currently set at 0.3 mg/kg body weight per day (EPA 1998), although this figure has been criticized as being too stringent, because it does not account for the bioavailability of zinc.

To determine the appropriate level of fortification in a particular situation, it is necessary to measure or estimate the amount of the food vehicle that is consumed by different segments of the population. If, for example, the aim is to provide young children with approximately half the recommended daily allowance of 10 mg of zinc (i.e., 5 mg) through a fortified cereal flour, the flour should be fortified at a level of 100 ppm (or 10 mg/100 g of flour) if the children consume on average 50 g of flour per day. Assuming that the maximum amount of flour that an adult will consume is less than 80% of total energy intake and also assuming that the upper range of energy intake is about 3000 kcal/d, then the upper extreme of daily flour consumption would be about 2400 kcal, or about 600 g of flour. At the aforementioned level of fortification, these high consumers of flour would receive 60 mg of zinc per day from the fortified cereal, a level that might be considered excessive.

Alternatively, the upper level of acceptable zinc intake from the fortified food might be fixed at 40 mg/d, which is considerably less than the upper level of intake for adults recommended by WHO (1996). In this case, adults who consume flour at the upper extreme of the expected distribution (i.e., 600 g of flour per day) could receive flour fortified at a maximum level of 6.7 mg zinc per 100 g of flour (67 ppm). At this level of fortification, a young child who consumes 50 grams of flour per day would receive an additional 3.3 mg of zinc. These estimates for appropriate levels of fortification should be reevaluated in local settings, using empirical information on dietary intake, where possible.

For the most part, the only items that are currently fortified with zinc in industrialized countries are special-purpose foods that reach targeted consumers, such as infant formulas and ready-to-eat breakfast cereals. A possible practical limitation in implementing universal zinc fortification programs in developing countries may result from the fact that, unlike the situation for many other nutrients, there are no programs now in place in industrialized countries to fortify staple foods or condiments with zinc. Although this is not surprising, given that zinc deficiency is unlikely to be a major public health problem in the more affluent countries, one consequence is that there is very little prior experience with large-scale zinc fortification programs. Another related issue is that some zinc compounds that could potentially be used as fortificants, such as zinc acetate, have not been added to the US food supply and therefore have not been evaluated for inclusion in the list of compounds that are generally recognized as safe (GRAS) by the US Food and Drug Administration (FDA) (National Archives and Records Administration 1999). It might be more difficult for some countries to consider using these compounds as fortificants without the GRAS designation.

Of the GRAS compounds, zinc sulfate and zinc oxide are the ones that are most attractive for universal fortification programs because of their relatively low cost compared with other zinc compounds. Although zinc oxide is considerably cheaper and more stable than zinc sulfate, the latter compound is preferable because of its presumed greater absorption, even by individuals who are hypochlorhydric. Nevertheless, questions have been raised regarding the possibility that zinc sulfate is actually converted to zinc oxide during cooking. If this is true, there would be no justification for using the more expensive compound. Obviously, this question needs to be resolved to provide adequate guidance to fortification programs. It has been estimated in Indonesia that fortification of wheat flour with zinc sulfate at a level of 30 ppm of zinc would cost about $0.47 per metric ton of flour (Ranum, personal communication[1]), a cost that is deemed affordable for most countries.

In countries where cereal flour is already being fortified with iron, the cost of zinc fortification would be similar to that of iron fortification. A practical advantage of the combined program is that the fortificants could be added simultaneously as a premix, and the adequacy of fortification could be monitored by measuring just one of the two elements.

---

[1] P. Ranum, American Ingredients Company, Kansas City, MO, USA, personal communication.

Conclusions of Background Review

On the basis of the information presented above concerning the amount of zinc available in national food supplies, it appears that the risk of low dietary intake of absorbable zinc and consequent zinc deficiency are widespread problems affecting between one-third and one-half of the world's population. Zinc deficiency may induce a number of critical functional abnormalities, including impaired reproductive performance, depressed immune function and secondary increases in the incidence and severity of infections, growth failure and secondary stunting, and abnormalities of neurobehavioural development. Because of the likely high global prevalence of zinc deficiency and the serious range of complications that can be induced by this condition, public health programs are urgently needed to prevent low zinc intake and poor absorption of zinc.

Primary nutrition-related options for program interventions to reduce the global prevalence of zinc deficiency include promotion of increased dietary intake of zinc-rich foods and dietary modifications to enhance zinc absorption, supplementation, and fortification. These program options are not listed in any particular order of priority; selection of specific actions will depend on local context, such as food availability, presence of a suitable vehicle for fortification, and existence of other nutrition programs. To facilitate the implementation of zinc intervention programs, information is still needed on which individuals and population subgroups are at greatest risk of deficiency and are thus most likely to benefit from intervention. Likewise, practical experience is needed to determine the best strategies for delivering additional zinc to these high-risk groups. With regard to zinc supplementation and fortification, additional information is needed on the most appropriate chemical form of zinc, the optimal dosing schedule, and any possible risks involved. Information is also needed on ways of incorporating zinc into ongoing micronutrient programs, both to facilitate the rapid initiation of these programs and to ensure the ultimate efficiency of their operation.

Discussion Groups

Charge to Groups

Each conference participant was assigned to one of three working groups to consider whether, given the current state of knowledge, specific actions can be recommended to initiate intervention programs to reduce the global prevalence of zinc deficiency and, if so, what those actions might be. The first group focused on dietary diversification or modification, the second group addressed zinc supplementation, and the third considered zinc fortification. Each group was asked a series of questions: Is currently available information sufficient to start recommending implementation of the specific type of intervention program? If so, what actions should be taken by individual countries and international agencies? If not, what additional information is needed to enable program development? In addition, each group was asked to consider ways of identifying target populations for intervention and feasible methods of monitoring program activities. The participants in each working group are listed in Appendix 2.

General Conclusions

There was strong consensus within each of the working groups that, even though many research questions remain to be addressed, public health programs should be implemented at the earliest possible opportunity to reduce the risk of zinc deficiency in countries where this problem is likely to occur. High risk of zinc deficiency is primarily a problem of low-income countries, so intervention programs are needed most urgently in these settings. When possible, zinc intervention programs should be carried out in the context of ongoing primary health care or existing nutrition programs, rather than through newly established, stand-alone programs to control zinc deficiency.

Thus, policymakers responsible for general health and nutrition programs should attempt to identify opportunities to integrate zinc interventions with their existing activities. For example, programs to promote improved complementary feeding should examine the local diet and feeding recommendations from the perspective of zinc nutriture and determine whether specific zinc-rich foods or food-processing techniques to enhance zinc absorption should be included in current feeding recommendations. Likewise, ongoing mineral supplementation or fortification programs, such as iron supplementation of pregnant women and iron fortification of wheat flour, should be reviewed to assess whether they should include zinc. Finally, opportunities to improve zinc status should also be sought within disease control programs. For example, zinc supplementation should be considered as adjunctive therapy for children being treated for acute or persistent diarrhea.

Most countries will need to assess the population's zinc status. Probably the best approach initially is to review — or collect, if necessary — representative data on dietary intake by high-risk segments of the population, especially children and women of reproductive age. Information is available on how to conduct and interpret such surveys (Murphy et al. 1992; Gibson and Ferguson 1998a). Because animal products are the best sources of readily absorbable zinc, intake is more likely to be inadequate in people who consume few or no animal products, whether for reasons of economic constraint or religious or cultural preference.

Information on the amount of total and absorbable zinc in the national food supply can also be used to determine the likelihood that the population may be at risk of inadequate zinc intake. The authors of the present document have reviewed the total and absorbable zinc in national food supplies and the percentage of the population at risk of low intake for almost all countries that have FAO food balance sheets (FAO 1998). This information is available upon request.

Another approach to assess population zinc status is to measure the population's mean serum or plasma zinc concentration. This necessitates proper collection and processing of blood from a representative sample of the population and appropriate analysis of the zinc concentration in the plasma or serum. Although there is not yet consensus on the interpretation of these values, there is some evidence to suggest that groups of children with mean serum zinc concentration less than 12.2 µmol/L (80 µg/dL) are likely to respond to zinc supplementation.

Another indirect approach for estimating the global prevalence of zinc deficiency in preschool children could take advantage of existing information on the prevalence of stunting (low height-for-age) and underweight (low weight-for-age), which is available from the United Nations agencies. There is fairly strong evidence that populations of children with significant degrees of stunting or underweight (mean height-for-age or weight-for-age Z-score less than -2) may grow more rapidly following zinc supplementation. Thus, it is reasonable to assume that excessive rates of stunting can serve as contributory information to suggest that zinc deficiency may exist in a group of preschool children.

Once a target population has been identified, the specific method (or methods) of intervention that is (are) most suited to that population must be determined. As just indicated, these interventions should be integrated into other ongoing programs to the extent possible. Finally, because of limited experience with large-scale zinc intervention programs, it would be prudent to begin with small-scale pilot projects to establish population-specific activities that are feasible and sustainable before moving to general implementation.

For the most part, general approaches to evaluate the success of zinc intervention programs are no different from those used for any type of nutrition intervention. In particular, evaluations should monitor both process indicators, such as the number of caregivers successfully trained in household food-processing techniques or the number of low-birth-weight infants who received zinc supplements, and indicators of the impact of the program on the target population's zinc status. Monitoring the impact of a zinc intervention program is perhaps somewhat more difficult than might be the case for other nutrients, because of the lack of suitable biomarkers of zinc status. Nevertheless, the working groups agreed that change in the mean serum zinc concentration of the target population would provide useful evidence that the intervention was affecting the population's zinc status. There is a clear need for development of simple, low-cost, reliable techniques to assess zinc status, both to determine the need for zinc intervention programs in particular population groups and to evaluate the impact of such programs.

Dietary Diversification or Modification

The working group on dietary diversification or modification defined the range of possible interventions in this category as changes in plant breeding and cultivation practices that increase the zinc content or bioavailability of plant-derived foods, promotion of increased consumption of zinc-rich foods, and reduction of intake of inhibitors of zinc absorption (by means of industrial or household-level processing and preparation of indigenous foods). The ultimate goals of these strategies are to increase access to and utilization of foods with a high content and bioavailability of zinc throughout the year. The group stressed that these interventions are important for preventing zinc deficiency and that they can be implemented using current knowledge. However, these approaches may not increase zinc intake and absorption sufficiently to cure preexisting zinc deficiency. Thus, other curative approaches, such as supplementation, may be necessary when zinc deficiency already exists.

The two general strategies to improve zinc utilization through dietary diversification or modification are by increasing the zinc content of the diet or enhancing its absorption. In both cases, a cornerstone of dietary strategies for young infants is the promotion and protection of exclusive breast-feeding for about the first 6 months of life. Strategies to increase the zinc content of the diet are particularly important when the traditional diet is based on staples that are generally low in zinc, such as starchy roots and tubers, for example, yam, cassava, potato, and sago. Polished rice has a lower zinc content than other cereals, so diets based on rice are also included in this category. Possible methods to increase the zinc content of these diets include cultivation techniques, such as use of zinc-containing fertilizers, and plant breeding or genetic engineering techniques to increase the zinc content of the staple food or foods. Whereas zinc-containing fertilizers are already used in agricultural programs, plant breeding and genetic engineering to enhance the zinc content of plant foods are areas for additional research.

At the household level, zinc intake could be enhanced by increased use of a greater range of zinc-rich foods, such as flesh foods, local plants that have high zinc contents, and possibly indigenous insects and grubs, where these are culturally acceptable. More information is needed on the zinc and phytate contents of local plants to identify those that might be suitable sources of absorbable zinc. Methods to increase consumption of animal products that are not currently consumed, such as organ meats or dried fish, might include incorporation of these items in processed foods, such as noodles, chips, or other snack foods.

Possible approaches to reduce the phytate content of the diet are through plant breeding and genetic engineering, food-processing and preparation techniques, and possibly addition of exogenous phytase to the diet. Selective breeding and genetic engineering research have produced strains of cereals with reduced phytate contents. One low-phytate strain of maize has already been demonstrated to enhance iron absorption (Mendoza et al. 1998), and could reasonably be expected to also enhance absorption of zinc. Although these strains are not yet in field production, they should become available in the not-distant future. Through genetic engineering it is also possible to incorporate phytase into cereal staples, although the effects of these genetically modified strains on the final phytate content of prepared foods and zinc absorption are not yet known.

At the household level, soaking of beans, maize, and possibly other cereals can reduce their phytate content substantially. Germination or fermentation of cereals also reduces their phytate content, although these processes may be acceptable only in cultures where they are traditionally practiced. Yeast treatment of dough reduces the phytate content of leavened breads, although recently developed "quick-rising" doughs reduce the amount of time that phytase-containing yeast can react with phytate to produce this effect. It might be possible to add industrially produced phytase to bread dough and other cereal and legume products, but as yet there is no practical experience with this approach in humans. It is worth noting that reduction in the phytate content of the diet would also favour enhanced absorption of other minerals, such as iron and calcium.

One aspect of household-level intervention that deserves mention is the critical importance of education and communication to foster any necessary change in behaviour. A considerable amount of background research on knowledge, attitudes, and practices with regard to food selection are needed before educational messages and materials to disseminate this information are prepared. A complete discussion of these issues is beyond the scope of the current document, and the reader is referred to other presentations of this material (Manoff 1985; Hornik 1988; Dickin et al. 1997).

Methods of monitoring dietary interventions may differ to some extent from those that are used to assess other forms of interventions. In particular, because of the prerequisite that food preparation or eating behaviour must change before proposed dietary interventions can have any nutritional impact, it is possible to assess the population's knowledge of the recommended behaviours and their actual food preparation practices and dietary intake to monitor the early effects of the program. Of course, the ultimate objective of the intervention is to improve nutritional status, as measured directly by assessing the population's mean plasma zinc concentration and possibly other functional indicators of zinc status, such as rates of nutritional stunting or underweight, but these more distal outcomes could not be expected to change in response to dietary interventions without the food-related behaviours changing first.

Supplementation

The discussion group on zinc supplementation concluded that, on the basis of current knowledge, interventions can be initiated in certain high-risk target groups, although additional research is needed on the optimal dose of supplements for different situations and on the operational aspects of formulating and distributing these supplements. The presently agreed upon primary target groups for supplementation are pregnant women, young children from 6 to about 24 months of age, and younger infants with low birth weight, because of both the relatively high zinc needs of these groups and the difficulty in reaching very young children with adequate amounts of fortified foods. Other high-risk groups that might be evaluated are older preschool and school-age children, patients with diarrheal disease (who have excessive fecal zinc excretion), adolescents, lactating women, and the elderly (who may have impaired zinc absorption). Refugee populations may also benefit from universal supplementation if the absorbable zinc content of their food supply is inadequate. Supplements may be used either for prevention of zinc deficiency in populations with a high risk of poor zinc status or for treatment of a particular condition, such as acute or persistent diarrhea, severe malnutrition (stunting or underweight), or intrauterine growth retardation.

As indicated above, the appropriate doses of supplemental zinc for various age groups and for various clinical conditions have not been studied systematically. Therefore, the discussion group proposed tentative recommendations based on the midrange of doses that have been used successfully in published clinical trials. The following dosages of zinc supplements were suggested for particular age groups and clinical situations: newborns, 2–4 mg/d; young children, 5–10 mg/d; young children with diarrhea or severe malnutrition, 10–20 mg/d; and pregnant women, 20–25 mg/d. These recommendations may need to be modified upward if the supplements are given with foods that inhibit absorption of zinc, but specific information is currently lacking. It was further suggested that to avoid any possible adverse effect of zinc supplementation on copper status, the zinc–copper molar ratio in the supplements should be about 10 : 1, up to a maximum of 1 mg of copper per day. Finally, the zinc–iron molar ratio in a combined supplement should be approximately 1 : 1 and should not exceed 2 : 1. All of these tentatively suggested doses should be reexamined in future controlled dose–response studies, with regard to both the efficacy of the zinc supplement and any possible toxic effects.

Several chemical forms of zinc can be used in supplements. The choice of a particular chemical form should be based on its solubility in water, taste, cost, and safety. Most published studies have used zinc sulfate, zinc gluconate, or zinc acetate, all of which are water soluble, or zinc oxide, which is less soluble in water. Water-soluble compounds are preferable because they are absorbed more efficiently. Because zinc acetate has not been evaluated by the FDA with regard to its GRAS status, the agency should be asked to review this compound. There is very little information on whether the functional impact of zinc supplements might be modified by inclusion of other nutrients in a multiple-micronutrient supplement, so additional evaluations are needed.

The optimal physical form of the supplement depends on the age of the target group, cultural preferences, and the possible desirability or need to include additional nutrients in the supplement. Young children need a liquid preparation or one that can be made into a liquid in the household or added directly to foods, such as the "sprinkles" referred to above. Absorption of zinc is likely to be less efficient when the supplements are given with foods, especially if these foods contain phytate. In contrast, dry supplements (tablets, capsules, or powders) are cheaper and more stable and permit inclusion of a broader range of nutrients. Recent experience with high-fat, micronutrient-fortified spreads suggest that these may provide another option for supplementation programs, although more experience is needed to assess their acceptability, efficacy, and impact on consumption of other components of the diet.

Most supplementation trials have provided zinc on a daily basis. According to current knowledge of zinc metabolism, it is reasonable to expect that supplements would have to be provided frequently to have the desired continuous impact on zinc status. However, a limited number of studies in which zinc was given either once weekly for an extended period or daily for just 2–4 weeks found that these forms of intermittent supplementation had significant positive impacts on functional outcomes, such as physical growth or rate of infections. Thus, the functional impact of temporary improvement of zinc status may last longer than the period during which the rapidly exchangeable pool of zinc is expanded. Additional information is needed on the impact, acceptability, safety, and cost of different dosing schedules.

The advisability of coupling zinc interventions with existing nutrition and health intervention programs has already been discussed. However, there are several situations in which the use of zinc supplements in ongoing programs might be particularly attractive. For example, use of zinc supplements as an adjunct to other forms of treatment of childhood diarrhea could be recommended for high-risk populations, because of the beneficial impact of zinc on duration of diarrhea and because fecal zinc excretion is excessive during diarrhea. Although the group considered the possibility of adding zinc to oral rehydration solutions, this idea was discarded because the wide range in consumption of these solutions would make it practically impossible to adequately control the dose of zinc. As another example, recent efforts to distribute vitamin A supplements at the time of routine childhood immunizations might be linked with simultaneous distribution of a short course of zinc supplements. This approach would be especially attractive if further studies confirm the value of intermittent zinc supplementation.

General issues of monitoring and evaluation of zinc intervention programs have already been described. With regard to zinc supplementation, a proximate indicator of program success might be consumption (or "disappearance") of the supplements. Because in some cases supplements will include other nutrients, it might be preferable to monitor the impact of the program on the status of these other nutrients, if they are easier to assess than zinc. Thus, for example, successful distribution of an iron–zinc supplement could be confirmed by monitoring its impact on the population's median serum ferritin concentration, if that concentration is initially low.

Fortification

As with the other groups, the working group on fortification concluded that there is sufficient information to justify initiation of zinc fortification programs, particularly where there is a high risk of zinc deficiency and where iron fortification programs are already being implemented or contemplated (which would provide an opportunity to couple zinc fortification with iron fortification). General guidelines for food fortification should be reviewed and adapted, as necessary, with regard to zinc. Primary target groups for zinc fortification programs are those who have elevated zinc needs, as described above, and who are likely to consume the available food vehicles. Possible vehicles for universal fortification programs are staple foods, such as cereals and cereal flours, and condiments, such as salt, sauces, and monosodium glutamate. Special-purpose foods, such as centrally processed complementary foods and other processed foods and beverages, can also be fortified with zinc to reach particular target populations. Given current efforts to introduce iron fortification of wheat and maize flour in many countries, the same staples could become prime vehicles for zinc fortification as well. However, because of concerns regarding the efficiency of zinc absorption from high-phytate foods, it would be preferable to select a food vehicle that is low in phytate. For example, zinc would be better absorbed from white flour than from whole-wheat flour. In situations where the fortified foods are high in phytate (phytate–zinc molar ratio > 15), fortification interventions should be accompanied by advice on methods to reduce the phytate content of the final product as served. If these recommendations involve soaking the raw product, it will be important to determine whether the zinc fortificant is lost along with phytate during soaking.

Considerations regarding the proper level of zinc to include in fortified products were presented in the background section of this report. The fortification discussion group concluded that the appropriate levels of zinc fortification for cereal staples should be between about 30 and 70 mg/kg, although the specific level would need to be determined locally in individual countries, on the basis of the existing deficit in dietary intake, the range of consumption of the proposed food vehicle, and a review of its phytate content. The upper levels of iron fortification of cereal flours are typically about 20–30 ppm because higher amounts of iron impart undesirable organoleptic characteristics. Because of concerns that a zinc–iron molar ratio greater than about 2 : 1

may depress iron absorption, the level of iron in the doubly fortified food may limit the level of zinc that can be included.

The chemical form of zinc to be selected for fortification depends on cost, organoleptic properties, and bioavailability. As discussed for zinc supplements, water-soluble zinc fortificants are preferable to ensure adequate zinc absorption, although the possibility that the chemical form of zinc changes during cooking or baking must be explored. Available information from a limited number of studies and experience with commercial products suggest that recommended levels of zinc fortification are not likely to affect the sensory properties or stability of likely food vehicles, although this should be checked for each new food vehicle.

In general, the methods of evaluating the impact of a zinc fortification program are similar to those discussed for other types of interventions. Program coverage can also be assessed in terms of the consumption of the fortified food, on the basis of dietary intake studies. Because zinc will likely be added to a food as a premix along with other nutrients, it may be necessary to monitor only one of these nutrients in the food product to confirm that an adequate level of the fortificant has been included.

Research Needs

Each of the working groups prepared a list of additional research studies that should be conducted to support program implementation. Most of these issues have been noted and explained in the foregoing sections, so they are simply listed in the following summary of research priorities.

1.   Development and evaluation of simple, rapid, low-cost, reliable methods to assess the zinc status of individuals and populations. Additional population groups that should be evaluated with regard to their zinc status and functional response to enhanced zinc intake or absorption are pregnant women (especially in the periconceptional period), lactating women, school-age children, adolescents, and the elderly.

2.   Measurement of zinc and phytate content of local foods, including indigenous plants.

3.   Determination of the effect of simple household-level food-processing techniques to reduce the phytate content of the major staple foods.

4.   Evaluation of the effect of exogenous phytase on the phytate content of foods.

5.   Selective breeding or genetic modification of major plant foods to increase zinc content or reduce phytate content.

6.   Efficacy trials of the impact of dietary diversification or modification interventions on zinc status and other functional outcomes.

7.   Studies of absorption of different chemical forms of zinc supplements provided with or without other micronutrients or foods with various levels of phytate. Studies are needed both in normal individuals and those with impaired secretion of gastric acid.

8. Dose–response studies of the efficacy and safety of zinc supplements provided with and without other micronutrients. Particular attention should be directed to evaluating the possible adverse effect of long-term, high-dose zinc supplementation on copper status. Studies of different dosing schedules, including intermittent doses, are also needed.

9. Determination of the optimal level of zinc fortification of foods in different settings and assessment of any possible adverse effects of zinc fortification on copper and iron status.

10. Analysis of the chemical form of zinc fortificants in various food vehicles before and after food processing and cooking, and the relationship between the chemical form and absorption of zinc from fortified foods and the organoleptic characteristics of these foods.

11. Efficacy trials of the impact of zinc fortification on zinc status and other functional outcomes.

12. Effectiveness studies and cost–benefit analyses of each of the foregoing intervention strategies.

Appendix 1. Conference Participants and Support Personnel

Nelson Almeida
Senior Scientist, Nutrition Research
W.K. Kellogg Institute
2 Hamblin Ave. East
Battle Creek, MI 49016-3232, USA
phone: (+1-616) 961-2149
fax: (+1-616) 961-3130
email: nelson.almeida@kellogg.com

France Bégin
Senior Program Specialist
The Micronutrient Initiative
PO Box 8500
Ottawa, ON K1G 3H9, Canada
phone: (+1-613) 236-6163 ext. 2490
fax: (+1-613) 236-9579
email: fbegin@idrc.ca

Robert E. Black
Edgar Berman Professor of International
    Health
The Johns Hopkins University School of
    Hygiene & Public Health
615 North Wolfe St.
Baltimore, MD 21205-2179, USA
phone: (+1-410) 955-3934
fax: (+1-410) 955-7159
email: rblack@jhsph.edu

André Briend
CNAM/ISTNA
5 rue du Vertbois
75003 Paris, France
phone: (+33-1) 53-01-80-36
fax: (+33-1) 53-01-80-05
email: brienda@cnam.fr

Kenneth H. Brown
Program in International Nutrition
Department of Nutrition
University of California, Davis
3150C Meyer Hall
Davis, CA 95616, USA
phone: (+1-530) 752-1992 ext. 0850 or
    (+1-530) 750-0159
fax: (+1-530) 752-3406
email: khbrown@ucdavis.edu

Ian Darnton-Hill
Vice President for Programs and Director of
    Nutrition and Health
Helen Keller International
90 West St., 2nd Floor
New York, NY 10006, USA
phone: (+1-212) 766-5266 ext. 824
fax: (+1-212) 791-7590
email: idarnton-hill@hki.org

Rosalind Gibson
Department of Human Nutrition
University of Otago
PO Box 56
Dunedin, Otago, New Zealand
phone: (+64-3) 479-7959 or (+64-3) 471-0929
fax: (+64-3) 479-7958
email: rosalind.gibson@
    stonebow.otago.ac.nz

K. Michael Hambidge
University of Colorado Health Sciences
    Center
4200 East Ninth Ave., C225
Denver, CO 80262, USA
phone: (+1-303) 315-5672
fax: (+1-303) 315-3273
email: Michael.Hambidge@UCHSC.edu

Janet C. King
Director and Professor
Western Human Nutrition Research Center
University of California, Davis
Davis, CA 95616, USA
phone: (+1-530) 752-5268
fax: (+1-530) 752-5271
email: jking@whnrc.usda.gov

Mahshid Lotfi
Senior Program Specialist
The Micronutrient Initiative
PO Box 8500
Ottawa, ON K1G 3H9, Canada
phone: (+1-613) 236-6163 ext. 2482
fax: (+1-613) 236-9579
email: mlotfi@idrc.ca

Venkatesh Mannar
Executive Director
The Micronutrient Initiative
PO Box 8500
Ottawa, ON K1G 3H9, Canada
phone: (+1-613) 236-6163 ext. 2210
fax: (+1-613) 236-9579
email: vmannar@idrc.ca

Haile Mehansho
Research Scientist
Miami Valley Laboratories
Proctor & Gamble Company
PO Box 398707
Cincinnati, OH 45239, USA
phone: (+1-513) 627-2725 or
    (+1-513) 983-1100 (head office)
fax: (+1-513) 627-1940
email: mehansho.h@pg.com

James Penland
Research Psychologist
Human Nutrition Research Centre
US Department of Agriculture
2420 Second Ave. North
PO Box 9034
Grand Forks, ND 58203-9034, USA
fax: (+1-701) 795-8220

Peter Ranum
2211 South Double O St.
Tuscon, AZ 85713, USA
fax: (+1-520) 883-7070
email: Doughmaker@aol.com

Brittmarie Sandström
Royal Veterinary and Agricultural University
Frederiksberg, Denmark

Noel W. Solomons
Scientific Co ordinator
Center for Studies of Sensory Impairment,
    Aging and Metabolism
PO Box 02 5339
Section 3163 – Guatemala
Miami, FL 33102-5339, USA
phone: (+1-502) 473-3942
fax: (+1-502) 473-3942
email: cessiam@guate.net

Stanley Zlotkin
Head, Division of GI/Nutrition
The Hospital for Sick Children
555 University Ave.
Toronto, ON M5G 1X8, Canada
phone: (+1-416) 813-6170
fax: (+1-416) 813-4972
email: zlotkin@sickkids.on.ca

Other contributors from University of California, Davis: faculty — Lindsay Allen, Kay Dewey, Carl Keen, William Lacy, Bo Lönnerdal, Robert Zierenberg; support personnel — D'Ann Finley, Marjorie Haskell, Daniel Lopez de Romaña, Diane Vandepeute, Sara E. Wuehler.

Appendix 2. Meeting Program

Thursday, October 21

1430   Registration, coffee

1500   Welcome – Kenneth H. Brown, Director, Program in International Nutrition, University of California, Davis; William Lacy, Vice Provost, University Outreach and International Programs, University of California, Davis

1510   Purpose of conference — Venkatesh Mannar, Executive Director, The Micronutrient Initiative

1520   History of zinc research at University of California, Davis — Carl Keen, Chair, Department of Nutrition, University of California, Davis

Session 1 — Chair: Janet King

1530   Overview and estimation of the global prevalence of zinc deficiency — Kenneth H. Brown, University of California, Davis

1615   Review of zinc metabolism and requirements; zinc and growth — K. Michael Hambidge, University of Colorado

1700   Dietary sources of zinc and factors affecting bioavailability — Noel W. Solomons, CESSIAM (Center for Studies of Sensory Impairment, Aging and Metabolism)

1745   Diagnosis of zinc deficiency in individuals and populations; zinc toxicity — Brittmarie Sandström, Royal Veterinary and Agricultural University

1830   Reception

1900   Dinner
        Speaker: Prof. Robert Zierenberg, Department of Geology, University of California, Davis

Friday, October 22

Session 2 — Chair: Ian Darnton-Hill

0900   Zinc and pregnancy outcome — Janet C. King, Western Human Nutrition Research Center, University of California, Davis

0945   Zinc and immune function, morbidity, mortality — Robert E. Black, The Johns Hopkins University

1030   Coffee

1100   Zinc and neurobehavioural function — James Penland, US Department of Agriculture Human Nutrition Research Center

Session 3 — Chair: Robert E. Black

1145   General experience with dietary modification — possible strategies (increased intake of zinc-rich foods, decreased intake of inhibitors of absorption), acceptability, impact — Rosalind Gibson, University of Otago

1245   Lunch

1400   General experience with zinc supplementation: forms of delivery, dose, acceptability, absorption, side effects, impact on zinc status (possible linkage with iron supplementation) — André Briend, CNAM/ISTNA; Stanley Zlotkin, Hospital for Sick Children

1530   General experience with zinc fortification — zinc compounds, level of fortification, possible food vehicles, effects on sensory properties of food, acceptability, absorption, side effects, impact on zinc status (possible linkage with iron fortification) — Peter Ranum, SUSTAIN (Sharing United States Technology to Aid in the Improvement of Nutrition); Nelson Almeida, W.K. Kellogg Institute; Haile Mehansho, Proctor & Gamble Company

1630   Working groups — Basic questions: Are we ready to start recommending intervention programs? If so, what? If not, what additional information is needed?

Group 1: Dietary modification

Issues: Identification of target group(s), appropriate and feasible food sources of zinc, appropriate and feasible processing techniques, dissemination of information, monitoring of program

Members of discussion group: Gibson (discussion leader), Hambidge, Lönnerdal, Lotfi, Sandström, Solomons

Group 2: Zinc supplementation

Issues: Identification of target group(s), dose (amount, frequency), form of delivery (chemical, physical), interaction with other nutrients, compatibility with other supplementation programs and other general intervention programs, cost, risk and toxicity, monitoring and evaluation

Members of discussion group: King (discussion leader), Allen, Bégin, Black, Briend, Penland

Group 3: Zinc fortification

Issues: Identification of target group(s), possible vehicles for fortification, dose (amount, frequency), form of delivery (chemical, physical), compatibility with other fortificants, possible effects on sensory properties and stability of food, monitoring of level in food, cost, risk and toxicity, evaluation

Members of discussion group: Darnton-Hill (discussion leader), Almeida, Brown, Mannar, Mehansho, Ranum, Zlotkin

Saturday, October 23

0900   Continuation of working groups

1030   Coffee

1100   Plenary reviews of working group discussions

1300   Lunch

1400   Continuation of discussion

Preparation of written report of meeting conclusions

References

Aggett, P.J. 1989. Severe zinc deficiency. *In* Mills, C.F., ed., Human nutrition reviews: zinc in human biology. Springer-Verlag, New York, NY, USA. pp. 259–280.

Aggett, P.J.; Favier, A. 1993. Zinc. International Journal for Vitamin and Nutrition Research, 63, 301–307.

Apgar, J. 1985. Zinc and reproduction. Annual Review of Nutrition, 5, 43–68.

Ashworth, A.; Morris, S.S.; Lira, P.I.C.; Grantham-McGregor, S.M. 1998. Zinc supplementation, mental development and behaviour in low birth weight term infants in northeast Brazil. European Journal of Clinical Nutrition, 52, 223–227.

Baer, M.T.; King, J.C. 1984. Tissue zinc levels and zinc excretion during experimental zinc depletion in young men. American Journal of Clinical Nutrition, 39, 556–570.

Bates, C.J.; Evans, P.H.; Dardenne, M.; Prentice, A.; Lunn, P.G.; Northrop-Clewes, C.A.; Hoare, S.; Cole, T.J.; Horan, S.J.; Longman, S.C.; Stirling, D.; Aggett, P.J. 1993. A trial of zinc supplementation in young rural Gambian children. British Journal of Nutrition, 69, 243–255.

Beisel, W.R.; Pekarek, R.S.; Wannemacher, R.W., Jr. 1973. The impact of infectious disease on trace-element metabolism of the host. *In* Hoekstra, W.G.; Suttie, J.; Ganther, H.; Mertz, W., ed., Trace element metabolism in animals, vol. 2. University Park Press, Baltimore, MD, USA. pp. 217–240.

Bentley, M.E.; Caulfield, L.E.; Ram, M.; Santizo, M.C.; Hurtado, E.; Rivera, J.A.; Ruel, M.T.; Brown, K.H. 1997. Zinc supplementation affects the activity patterns of rural Guatemalan infants. Journal of Nutrition, 127, 1333–1338.

Bhutta, Z.A.; Black, R.E.; Brown, K.H.; Meeks Gardner, J.; Gore, S.; Hidayat, A.; Khatun, F.; Martorell, R.; Ninh, N.X.; Penny, M.E.; Rosado, J.L.; Roy, S.K.; Ruel, M.; Sazawal, S.; Shankar, A. 1999. Prevention of diarrhea and pneumonia by zinc supplementation in children in developing countries: pooled analysis of randomized controlled trials. Zinc Investigators' Collaborative Group. Journal of Pediatrics, 135, 689–697.

Bhutta, Z.A.; Nizami, S.Q.; Isani, Z. 1999. Zinc supplementation in malnourished children with persistent diarrhea in Pakistan. Pediatrics, 103(4), e42. Available from http://www.pediatrics.org/cgi/content/full/103/4/e42. Cited 25 July 2000.

Black, M. 1998. Zinc deficiency and child development. American Journal of Clinical Nutrition, 68(Suppl), 464S–469S.

Briend, A.; Lascala, R.; Prudhon, C.; Mournier, B.; Grellety, Y.; Golden, M.H.N. 1999. Ready-to-use therapeutic food for treatment of marasmus. Lancet, 353, 1767–1768.

Brown, K.H. 1998. Effect of infections on plasma zinc concentration and implications for zinc status assessment in low-income countries. American Journal of Clinical Nutrition, 68(Suppl), 425S–439S.

Brown, K.H.; Peerson, J.M.; Allen, L.H. 1998. Effect of zinc supplementation on children's growth: a meta-analysis of intervention trials. Bibliotheca Nutritio et Dieta, 54, 76–83.

Budavari, S.; O'Neil, M.J.; Smith, A.; Heckleman, P.E.; Kinneary, J.F., ed. 1996. The Merck index: an encyclopedia of chemicals, drugs, and biologicals (12th ed.). Merck Research Laboratories, Division of Merck & Co., Inc., Whitehouse Station, NJ, USA.

Bunch, S.; Murphy, S.P. 1994. User's guide to the operation of the World Food Dietary Assessment Program. Office of Technology Licensing, University of California, Berkeley, CA, USA. 21 pp.

Castillo-Duran, C.; Garcia, H.; Venegas, P.; Torrealba, I.; Panteon, E.; Concha, N.; Perez, P. 1994. Zinc supplementation increases growth velocity of male children and adolescents with short stature. Acta Paediatrica, 83, 833–837.

Castillo-Duran, C.; Heresi, G.; Fisberg, M.; Uauy, R. 1987. Controlled trial of zinc supplementation during recovery from malnutrition: effects on growth and immune function. American Journal of Clinical Nutrition, 45, 602–608.

Castillo-Duran, C.; Rodriguez, A.; Venegas, G.; Alvarez, P.; Icaza, G. 1995. Zinc supplementation and growth of infants born small for gestational age. Journal of Pediatrics, 127, 206–211.

Castillo-Duran, C.; Vial, P.; Uauy, R. 1988. Trace mineral balance during acute diarrhea in infants. Journal of Pediatrics, 113, 452–457.

Caulfield, L.E.; Zavaleta, N.; Figueroa, A.; Leon, Z. 1999. Maternal zinc supplementation does not affect size at birth or pregnancy duration in Peru. Journal of Nutrition, 129, 1563–1568.

Caulfield, L.E.; Zavaleta, N.; Shankar, A.H.; Merialdi, M. 1998. Potential contribution of maternal zinc supplementation during pregnancy to maternal and child survival. American Journal of Clinical Nutrition, 68(Suppl), 499S–508S.

Cavan, K.R.; Gibson, R.S.; Grazioso, C.F.; Isalgue, A.M.; Ruz, M.; Solomons, N.W. 1993. Growth and body composition of peri-urban Guatemalan children in relation to zinc status: a cross-sectional study. American Journal of Clinical Nutrition, 57, 334–343.

Chandra, R.K. 1984. Excessive intake of zinc impairs immune responses. Journal of the American Medical Association, 252, 1443–1446.

Cherry, F.F.; Sandstead, H.H.; Rojas, P.; Johnson, L.K.; Batson, K.H.; Wang, X.B. 1989. Adolescent pregnancy: associations among body weight, zinc nutriture, and pregnancy outcome. American Journal of Clinical Nutrition, 50, 945–954.

Cousins, R.J. 1989. Systemic transport of zinc. In Mills, C.F., ed., Human nutrition reviews: zinc in human biology. Springer-Verlag, New York, NY, USA. pp. 79–93.

———— 1996. Zinc. In Ziegler, E.E.; Filer, L.J., Jr, ed., Present knowledge in nutrition (7th ed.). International Life Sciences Institute Press, Washington, DC, USA. pp. 293–306.

Dale, A.; Thomas, J.E.; Darboe, M.K.; Coward, W.A.; Harding, M.; Weaver, L.T. 1998. Helicobacter pylori infection, gastric acid secretion, and infant growth. Journal of Pediatric Gastroenterology and Nutrition, 26, 393-397.

Davidsson, L. 1999. Zinc. In Hurrell, R., ed., The mineral fortification of foods. Leatherhead Publishing, Surrey, UK. pp. 187–197.

Dickin, K.; Griffiths, M.; Piwoz, E. 1997. Designing by dialogue: a program planners' guide to consultative research for improving young child feeding. Academy for Educational Development, Washington, DC, USA.

Dirren, H.; Barclay, D.; Ramos, J.G.; Lozano, R.; Montalvo, M.M.; Davila, N.; Mora, J.O. 1994. Zinc supplementation and child growth in Ecuador. Advances in Experimental Medicine and Biology, 352, 215–222.

EPA (Environmental Protection Agency). 1998. Integrated risk information system: zinc and compounds. CASRN 7440-66-6. Available from http://www.epa.gov/ngispgm3/iris/subst/0426.htm. Updated 5 May 1998; cited 20 Jan. 2000.

Falchuk, K.H. 1977. Effect of acute disease and ACTH on serum zinc proteins. New England Journal of Medicine, 296, 1129–1134.

FAO (Food and Agriculture Organization). 1998. Food balance sheets, country averages 1990–1997. In FAO statistical databases [databases online]. Available from http://apps.fao.org/lim500/nph-wrap.pl?FoodBalanceSheet&Domain=FoodBalanceSheet. Cited June–August 1999.

Farah, D.A.; Hall, M.J.; Mills, P.R.; Russell, R.I. 1984. Effect of wheat bran on zinc absorption. Human Nutrition Clinical Nutrition, 38(6), 433–441.

Fischer, P.W.F.; Giroux, A.; L'Abbé, M.R. 1984. Effect of zinc supplementation on copper status in adult man. American Journal of Clinical Nutrition, 40, 743–746.

Flanagan, P.R.; Cluett, J.; Chamberlain, M.J.; Valberg, L.S. 1985. Dual-isotope method for determination of human zinc absorption: the use of a test meal of turkey meat. Journal of Nutrition, 115, 112–122.

Friel, J.K.; Andrews, W.L.; Matthew, J.D.; Long, D.R.; Cornel, A.M.; Cox, M.; McKim, E.; Zerbe, G.O. 1993. Zinc supplementation in very-low-birth-weight infants. Journal of Pediatric Gastroenterology and Nutrition, 17, 97–104.

Friis, H.; Ndhlovu, P.; Mduluza, T.; Kaondera, K.; Sandström, B.; Michaelsen, K.F.; Vennervald, B.J.; Christensen, N.O. 1997. The impact of zinc supplementation on growth and body composition: a randomized, controlled trial among rural Zimbabwean schoolchildren. European Journal of Clinical Nutrition, 51, 38–45.

Garg, H.K.; Singhal, K.C.; Arshad, Z. 1993. A study of the effect of oral zinc supplementation during pregnancy on pregnancy outcome. Indian Journal of Physiology and Pharmacology, 37, 276–284.

Gatheru, Z.; Kinoti, S.; Alwar, J.; Mwita, M. 1988. Serum zinc levels in children with kwashiorkor aged one to three years at Kenyatta National Hospital and the effect of zinc supplementation during recovery. East African Medical Journal, 65, 670–679.

Gibson, R.S. 1990. Principles of nutritional assessment. Oxford University Press, New York, NY, USA . 691 pp.

———— 1994. Zinc nutrition in developing countries. Nutrition Research Reviews, 7, 151–173.

Gibson, R.S.; Ferguson, E.L. 1998a. Assessment of dietary zinc in a population. American Journal of Clinical Nutrition, 68(Suppl), 430S–434S.

———— 1998b. Nutrition intervention strategies to combat zinc deficiency in developing countries. Nutrition Research Reviews, 11, 115–131.

———— 1999. An interactive 24-hour recall for assessing the adequacy of iron and zinc intakes in developing countries. International Life Sciences Institute Press, Washington, DC, USA.

Gibson, R.S.; Smit Vanderkooy, P.D.; MacDonald, A.C.; Goldman, A.; Ryan, B.A.; Berry, M. 1989. A growth-limiting, mild zinc-deficiency syndrome in some Southern Ontario boys with low height percentiles. American Journal of Clinical Nutrition, 49, 1266–1273.

Gilman, R.H.; Partanen, R.; Brown, K.H.; Spira, W.M.; Khanam, S.; Greenberg, B.; Bloom, S.R.; Ali, A. 1988. Decreased gastric acid secretion and bacterial colonization of the stomach in severely malnourished Bangladeshi children. Gastroenterology, 94, 1308–1314.

Golden, B.E.; Golden, M.H.N. 1992. Effect of zinc on lean tissue synthesis during recovery from malnutrition. European Journal of Clinical Nutrition, 46, 697–706.

Goldenberg, R.L.; Tamura, T.; Neggers, Y.; Copper, R.L.; Johnston, K.E.; DuBard, M.B.; Hauth, J.C. 1995. The effect of zinc supplementation on pregnancy outcome. Journal of the American Medical Association, 274, 463–468.

Golub, M.S.; Gershwin, M.E.; Hurley, L.S.; Hendrickx, A.G.; Saito, W.Y. 1985. Studies of marginal zinc deprivation in rhesus monkeys: infant behavior. American Journal of Clinical Nutrition, 42, 1229–1239.

Golub, M.S.; Takeuchi, P.T.; Keen, C.L.; Gershwin, M.E.; Hendrickx, A.G.; Lönnerdal, B. 1994. Modulation of behavioral performance of prepubertal monkeys by moderate dietary zinc deprivation. American Journal of Clinical Nutrition, 60, 238–243.

Golub, M.S.; Takeuchi, P.T.; Keen, C.L.; Hendrickx, A.G.; Gershwin, M.E. 1996. Activity and attention in zinc-deprived adolescent monkeys. American Journal of Clinical Nutrition, 64, 908–915.

Hambidge, K.M.; Chavez, M.N.; Brown, R.M.; Walravens, P.A. 1978. Zinc supplementation of low-income pre-school children. *In* Kirchgessner, M., ed., Proceedings of the 3rd International Symposium on Trace Element Metabolism in Man and Animals. Institut für Ernahrungsphysiologie, Technische Universität Munchen, Freising-Weihenstephan, Germany. pp. 296–299.

———— 1979. Zinc nutritional status of young middle-income children and effects of consuming zinc-fortified breakfast cereals. American Journal of Clinical Nutrition, 32, 2532–2539.

Hambidge, K.M.; Goodall, M.J.; Stall, C.; Pritts, J. 1989. Post-prandial and daily changes in plasma zinc. Journal of Trace Elements and Electrolytes in Health and Disease, 3, 55–57.

Hambidge, M.; Krebs, N. 1995. Assessment of zinc status in man. Indian Journal of Pediatrics, 62, 157–168.

Heinig, M.J.; Brown, K.H.; Lönnerdal, B.; Dewey, K.G. 1998. Zinc supplementation does not affect growth, morbidity, or motor development of U.S. breastfed infants at 4–10 mo (abstract). FASEB Journal, 12, A970.

Henderson, L.M.; Brewer, G.J.; Dressman, J.B.; Swidan, S.Z.; DuRoss, D.J.; Adair, C.H.; Barnett, J.L.; Berardi, R.R. 1995. Effect of intragastric pH on the absorption of oral zinc acetate and zinc oxide in young healthy volunteers. Journal of Parenteral and Enteral Nutrition, 19, 393–397.

Henkin, R.I.; Patten, B.M.; Re, P.K.; Bronzert, D.A. 1975. A syndrome of acute zinc loss. Archives of Neurology, 32, 745–751.

Henry, R.W.; Elmes, M.E. 1975. Plasma zinc in acute starvation. British Medical Journal, 4(5997), 625–626.

Hooper, P.L.; Visconti, L.; Garry, P.J.; Johnson, G.E. 1980. Zinc lowers high-density lipoprotein-cholesterol levels. Journal of the American Medical Association, 244, 1960–1961.

Hornik, R. 1988. Development communication. Longman, New York, NY, USA. 182 pp.

References

Hunt, I.F.; Murphy, N.J.; Cleaver, A.E.; Faraji, B.; Swendseid, M.E.; Browdy, B.L.; Coulson, A.H.; Clark, V.A.; Settlage, R.H.; Smith, J.C., Jr. 1985. Zinc supplementation during pregnancy in low-income teenagers of Mexican descent: effects on selected blood constituents and on progress and outcome of pregnancy. American Journal of Clinical Nutrition, 42, 815–828.

Hunt, I.F.; Murphy, N.J.; Cleaver, A.E.; Faraji, B.; Swendseid, M.E.; Coulson, A.H.; Clark, V.A.; Browdy, B.L.; Cabalum, M.T.; Smith, J.C., Jr. 1984. Zinc supplementation during pregnancy: effects on selected blood constituents and on progress and outcome of pregnancy in low-income women of Mexican descent. American Journal of Clinical Nutrition, 40, 508–521.

Hurley, L.S.; Baly, D.L. 1982. The effects of zinc deficiency during pregnancy. In Prasad, A.S., ed., Clinical, biochemical, and nutritional aspects of trace elements. Liss, New York, NY, USA. pp. 145–159.

Hurrell, R., ed. 1999. The mineral fortification of foods. Leatherhead Publishing, Surrey, UK. 315 pp.

Jackson, M.J. 1989. Physiology of zinc: general aspects. In Mills, C.F., ed, Human nutrition reviews: zinc in human biology. Springer-Verlag, New York, NY, USA. pp. 1–14.

Jameson, S.; Ursing, I. 1976. Low serum zinc concentrations in pregnancy, results of investigations and treatment. Acta Medica Scandinavica Supplement, 593, 50–64.

Jønsson, B.; Hauge, B.; Larsen, M.F.; Hald, F. 1996. Zinc supplementation during pregnancy: a double blind randomised controlled trial. Acta Obstetricia et Gynecologica Scandinavica, 75, 725–729.

Keen, C.L.; Hurley, L.S. 1989. Zinc and reproduction: effects of deficiency on foetal and postnatal development. In Mills, C.F., ed., Human nutrition reviews: zinc in human biology. Springer-Verlag, New York, NY, USA. pp. 183–220.

Khanum, S.; Alam, A.N.; Anwar, I.; Akbar Ali, M.; Mujibur Rahaman, M. 1988. Effect of zinc supplementation on the dietary intake and weight gain of Bangladeshi children recovering from protein-energy malnutrition. European Journal of Clinical Nutrition, 42, 709–714.

Kilic, I.; Ozalp, I.; Coskun, T.; Tokatli, A.; Emre, S.; Saldamli, I.; Koksel, H.; Ozboy, O. 1998. The effect of zinc-supplemented bread consumption on school children with asymptomatic zinc deficiency. Journal of Pediatric Gastroenterology and Nutrition, 26, 167–171.

King, J.C. 1990. Assessment of zinc status. Journal of Nutrition, 120, 1474–1479.

Kynast, G.; Saling, E. 1986. Effect of oral zinc application during pregnancy. Gynecologic and Obstetric Investigation, 21, 117–123.

Lira, P.I.; Ashworth, A.; Morris, S.S. 1998. Effect of zinc supplementation on the morbidity, immune function, and growth of low-birth-weight, full-term infants in northeast Brazil. American Journal of Clinical Nutrition, 68, 418S–424S.

Lönnerdal, B.; Cederblad, Å.; Davidsson, L.; Sandström, B. 1984. The effect of individual components of soy formula and cows' milk formula on zinc bioavailability. American Journal of Clinical Nutrition, 40, 1064–1070.

Lönnerdal, B.; Sandberg, A.-S.; Sandström, B.; Kunz, C. 1989. Inhibitory effects of phytic acid and other inositol phosphates on zinc and calcium absorption in suckling rats. Journal of Nutrition, 119, 211–214.

Lukaski, H.C.; Bolonchuk, W.W.; Klevay, L.M.; Milne, D.B.; Sandstead, H.H. 1984. Changes in plasma zinc content after exercise in men fed a low-zinc diet. American Journal of Physiology, 247, E88–E93.

Mahomed, K.; James, D.K.; Golding, J.; McCabe, R. 1989. Zinc supplementation during pregnancy: a double blind randomised controlled trial. British Medical Journal, 299, 826–830.

Manoff, R.K. 1985. Social marketing: new imperatives for public health. Praeger, New York, NY, USA. 293 pp.

McKenzie, J.M.; Fosmire, G.J.; Sandstead, H.H. 1975. Zinc deficiency during the latter third of pregnancy: effects on fetal rat brain, liver and placenta. Journal of Nutrition, 105, 1466–1475.

Meeks Gardner, J.; Witter, M.M.; Ramdath, D.D. 1998. Zinc supplementation: effects on the growth and morbidity of undernourished Jamaican children. European Journal of Clinical Nutrition, 52, 34–39.

Mendoza, C.; Viteri, F.E.; Lönnerdal, B.; Young, K.A.; Raboy, V.; Brown, K.H. 1998. Effect of genetically modified low-phytate maize on absorption of iron from tortilla. American Journal of Clinical Nutrition, 68, 1123–1127.

Merialdi, M.; Caulfield, L.E.; Zavaleta, N.; Figueroa, A.; DiPietro, J.A. 1998. Adding zinc to prenatal iron and folate tablets improves fetal neurobehavioral development. American Journal of Obstetrics and Gynecology, 180, 483–490.

Moynahan, E.J. 1974. Acrodermatitis enteropathica: a lethal inherited human zinc-deficiency disorder. Lancet, 2, 399–400.

Murphy, S.; Beaton, G.H.; Calloway, D.H. 1992. Estimated mineral intakes of toddlers: predicted prevalence of inadequacy in village populations in Egypt, Kenya, and Mexico. American Journal of Clinical Nutrition, 56, 565–572.

Nakamura, T.; Nishiyama, S.; Futagoishi-Suginohara, Y.; Matsuda, I.; Higashi, A. 1993. Mild to moderate zinc deficiency in short children: effect of zinc supplementation on linear growth velocity. Journal of Pediatrics, 123, 65–69.

National Archives and Records Administration. 1999. Title 21, Food and drugs. Vol. 3, parts 182 and 582 (for zinc chloride, zinc gluconate, zinc oxide, zinc stearate, and zinc sulfate). Substances generally recognized as safe. In Code of Federal Regulations Online via GPO (Government Printing Office) Access. US Government Printing Office, Washington, DC, USA. Available from http://www.access.gpo.gov/nara/cfr/index.html. Updated 1 Apr. 1999, cited 7 Jan. 2000.

National Research Council. 1989. Recommended dietary allowances (10th ed.). National Academy Press, Washington, DC, USA. 302 pp.

Nävert, B.; Sandström, B.; Cederblad, Å. 1985. Reduction of the phytate content of bran by leavening in bread and its effect on zinc absorption in man. British Journal of Nutrition, 53, 47–53.

Ninh, N.X.; Thissen, J.P.; Collette, L.; Gerard, G.; Khoi, H.H.; Ketelslegers, J.M. 1996. Zinc supplementation increases growth and circulating insulin-like growth factor I (IGF-I) in growth-retarded Vietnamese children. American Journal of Clinical Nutrition, 63, 514–519.

Oberleas, D. 1983. Phytate content in cereals and legumes and methods of determination. Cereal Foods World, 28, 352–357.

Oberleas, D.; Harland, B.F. 1981. Phytate content of foods: effect on dietary zinc bioavailability. Journal of the American Dietetic Association, 79, 433–436.

O'Dell, B.L.; Reeves, P.G. 1989. Zinc status and food intake. In Mills, C.F., ed, Human nutrition reviews: zinc in human biology. Springer-Verlag, New York, NY, USA. pp. 173–181.

Olin, S.S. 1998. Between a rock and a hard place: methods for setting dietary allowances and exposure limits for essential minerals. Journal of Nutrition, 128, 364S–367S.

Osendarp, S.J.M.; Baqui, A.H.; Wahed, M.A.; Arifeen, S.E.; van Raaij, J.M.A.; Fuchs, G.J. 1998. Zinc supplementation during pregnancy in Bangladeshi women had no effect on birth weight (abstract 3763). FASEB Journal, 12, A647.

Panel on Dietary Reference Values of the Committee on Medical Aspects of Food Policy. 1991. Dietary reference values for food energy and nutrients for the United Kingdom. Her Majesty's Stationery Office, London, UK. 210 pp.

Patterson, W.P.; Winklemann, M.; Perry, M.C. 1985. Zinc-induced copper deficiency: megamineral sideroblastic anemia. Annals of Internal Medicine, 103, 385–386.

Penland, J.G.; Sandstead, H.H.; Alcock, N.W.; Dayal, H.H.; Chen, X.C.; Li, J.S.; Zhao, F.; Yang, J.J. 1997. A preliminary report: effects of zinc and micronutrient repletion on growth and neuropsychological function of urban Chinese children. Journal of the American College of Nutrition, 16, 268–272.

Penny, M.E.; Brown, K.H.; Lanata, C.L.; Peerson, J.M.; Marin, R.M.; Duran, A. 1997. Community-based trial of the effect of zinc supplements with and without other micronutrients on the duration of persistent diarrhea and the prevention of subsequent morbidity (abstract 3778). FASEB Journal, 11(3), A655.

Penny, M.E.; Peerson, J.M.; Marin, R.M.; Duran, A.; Lanata, C.F.; Lönnerdal, B.; Black, R.E.; Brown, K.H. 1999. Randomized, community-based trial of the effect of zinc supplementation, with and without other micronutrients, on the duration of persistent childhood diarrhea in Lima, Peru. Journal of Pediatrics, 135, 208–217.

Porter, K.G.; McMaster, D.; Elmes, M.E.; Love, A.H.G. 1977. Anaemia and low serum-copper during zinc therapy. Lancet, 2, 774.

Prasad, A.S. 1990. Discovery of human zinc deficiency and marginal deficiency of zinc. In Tomita, H., ed., Trace elements in clinical medicine. Springer-Verlag, Tokyo, Japan. pp. 3–11.

Prohaska, J.R.; Luecke, R.W.; Jasinski, R. 1974. Effect of zinc deficiency from day 18 of gestation and/or during lactation on the development of some rat brain enzymes. Journal of Nutrition, 104, 1525–1531.

Reddy, N.R.; Pierson, M.D.; Sathe, S.K.; Salunkhe, D.K. 1989. Phytates in cereals and legumes. CRC Press, Inc., Boca Raton, FL, USA. 152 pp.

Rivera, J.A.; Ruel, M.T.; Santizo, M.C.; Lönnerdal, B.; Brown, K.H. 1998. Zinc supplementation improves the growth of stunted rural Guatemalan infants. Journal of Nutrition, 128, 556–562.

Ronaghy, H.A.; Reinhold, J.G.; Mahloudji, M.; Ghavami, P.; Spivey Fox, M.R.; Halsted, J.A. 1974. Zinc supplementation of malnourished schoolboys in Iran: increased growth and other effects. American Journal of Clinical Nutrition, 27, 112–121.

Rosado, J.L.; Lopez, P.; Munoz, E.; Martinez, H.; Allen, L.H. 1997. Zinc supplementation reduced morbidity, but neither zinc nor iron supplementation affected growth or body composition of Mexican preschoolers. American Journal of Clinical Nutrition, 65, 13–19.

Ross, S.M.; Nel, E.; Naeye, R.L. 1985. Differing effects of low and high bulk maternal dietary supplements during pregnancy. Early Human Development, 10, 295–302.

Roy, S.K.; Tomkins, A.M.; Haider, R.; Behrens, R.H.; Akramuzzaman, S.M.; Mahalanabis, D.; Fuchs, G.J. 1999. Impact of zinc supplementation on subsequent growth and morbidity in Bangladeshi children with acute diarrhoea. European Journal of Clinical Nutrition, 53, 529–534.

Roy, S.K.; Tomkins, A.M.; Mahalanabis, D.; Akramuzzaman, S.M.; Haider, R.; Behrens, R.H.; Fuchs, G. 1998. Impact of zinc supplementation on persistent diarrhoea in malnourished Bangladeshi children. Acta Paediatrica, 87, 1235–1239.

Ruel, M.T.; Rivera, J.A.; Santizo, M.C.; Lönnerdal, B.; Brown, K.H. 1997. Impact of zinc supplementation on morbidity from diarrhea and respiratory infections among rural Guatemalan children. Pediatrics, 99, 808–813.

Ruz, M.; Castillo-Duran, C.; Lara, X.; Codoceo, J.; Rebolledo, A.; Atalah, E. 1997. A 14-mo zinc-supplementation trial in apparently healthy Chilean preschool children. American Journal of Clinical Nutrition, 66, 1406–1413.

Sandstead, H.H. 1991. Zinc deficiency. A public health problem? American Journal of Diseases of Childhood, 145, 853–859.

Sandstead, H.H.; Penland, J.G.; Alcock, N.W.; Dayal, H.H.; Chen, X.C.; Li, J.S.; Zhao, F.; Yang, J.J. 1998. Effects of repletion with zinc and other micronutrients on neuropsychologic performance and growth of Chinese children. American Journal of Clinical Nutrition, 68(Suppl), 470S-475S.

Sandström, B.; Almgren, A.; Kivistö, B.; Cederblad, Å. 1989. Effect of protein level and protein source on zinc absorption in humans. Journal of Nutrition, 119, 48–53.

Sandström, B.; Arvidsson, B.; Cederblad, Å.; Björn-Rasmussen, E. 1980. Zinc absorption from composite meals. I. The significance of wheat extraction rate, zinc, calcium, and protein content in meals based on bread. American Journal of Clinical Nutrition, 33, 739–745.

Sandström, B.; Cederblad, Å. 1980. Zinc absorption from composite meals. II. Influence of the main protein source. American Journal of Clinical Nutrition, 33, 1778–1783.

————— 1987. Effect of ascorbic acid on the absorption of zinc and calcium in man. International Journal of Vitamin and Nutrition Research, 57, 87–90.

Sandström, B.; Davidsson, L.; Cederblad, Å.; Lönnerdal, B. 1985. Oral iron, dietary ligands and zinc absorption. Journal of Nutrition, 115, 411–414.

Sandström, B.; Davidsson, L.; Eriksson, R.; Alpsten, M.; Bogentoft, C. 1987. Retention of selenium (^{75}Se), zinc (^{65}Zn) and manganese (^{54}Mn) in humans after intake of a labelled vitamin and mineral supplement. Journal of Trace Element and Electrolytes in Health and Disease, 1, 33–38.

Sazawal, S.; Bentley, M.; Black, R.E.; Dhingra, P.; George, S.; Bhan, M.K. 1996. Effect of zinc supplementation on observed activity in low socioeconomic Indian preschool children. Pediatrics, 98, 1132–1137.

Sazawal, S.; Black, R.E.; Bhan, M.K.; Bhandari, N.; Sinha, A.; Jalla, S. 1995. Zinc supplementation in young children with acute diarrhea in India. New England Journal of Medicine, 333, 839–844.

Sazawal, S.; Black, R.E.; Bhan, M.K.; Jalla, S.; Sinha, A.; Bhandari, N. 1997. Efficacy of zinc supplementation in reducing the incidence and prevalence of acute diarrhea — a community-based, double-blind, controlled trial. American Journal of Clinical Nutrition, 66, 413–418.

Sazawal, S.; Black, R.E.; Jalla, S.; Mazumdar, S.; Sinha, A.; Bhan, M.K. 1998. Zinc supplementation reduces the incidence of acute lower respiratory infections in infants and preschool children: a double-blind, controlled trial. Pediatrics, 102, 1–5.

Schlesinger, L.; Arevalo, M.; Arredondo, S.; Diaz, M.; Lönnerdal, B.; Stekel, A. 1992. Effect of a zinc-fortified formula on immunocompetence and growth of malnourished infants. American Journal of Clinical Nutrition, 56, 491–498.

Sempértegui, F.; Estrella, B.; Correa, E.; Aguirre, L.; Saa, B.; Torres, M.; Navarrete, F.; Alarcon, C.; Carrion, J.; Rodriguez Griffiths, J.K. 1996. Effects of short term zinc supplementation on cellular immunity, respiratory symptoms, and growth of malnourished Equadorian children. European Journal of Clinical Nutrition, 50, 42–46.

Shankar, A.H.; Genton, B.; Tamja, S.; Arnold, S.; Wu, L.; Baisor, M.; Paino, J.; Tielsch, J.A.; Alpers, M.A.; West, K.P., Jr. 1997. Zinc supplementation can reduce malaria-related morbidity in preschool children (abstract). American Journal of Tropical Medicine and Hygeine, 57, A434.

Shenkin, A. 1995. Trace elements and inflammatory response: implications for nutritional support. Nutrition, 11, 100–105.

Shrimpton, R. 1993. Zinc deficiency — is it widespread but under-recognized? ACC/SCN News, 9, 24–27.

Shrivastava, S.P.; Roy, A.K.; Jana, U.K. 1993. Zinc supplementation in protein energy malnutrition. Indian Pediatrics, 30, 779–782.

Simmer, K.; Khanum, S.; Carlsson, L.; Thompson, R.P.H. 1988. Nutritional rehabilitation in Bangladesh — the importance of zinc. American Journal of Clinical Nutrition, 47, 1036–1040.

Simmer, K.; Lort-Phillips, L.; James, C.; Thompson, R.P.H. 1991. A double-blind trial of zinc supplementation in pregnancy. European Journal of Clinical Nutrition, 45, 139–144.

Smith, K.T.; Failla, M.L.; Cousins, R.J. 1978. Identification of albumin as the plasma carrier for zinc absorption by perfused rat intestine. Biochemical Journal, 184, 627–633.

Smith, R.M.; King, R.A.; Spargo, R.M.; Cheek, D.B.; Field, J.B.;Veitch, L.G. 1985. Growth-retarded aboriginal children with low plasma zinc levels do not show a growth response to supplementary zinc. Lancet, 1, 923–924.

SUSTAIN (Sharing United States Technology to Aid in the Improvement of Nutrition). 1998. Forum on iron fortification. SUSTAIN, Washington, DC, USA. 54 pp.

Swanson, C.A.; King, J.C. 1983. Reduced serum zinc concentration during pregnancy. Obstetrics and Gynecology, 62, 313–318.

Tamura, T.; Goldenberg, R.L. 1996. Zinc nutriture and pregnancy outcome. Nutrition Research, 16, 139–181.

Thu, B.D.; Schultink, W.; Dillon, D.; Gross, R.; Leswara, N.D.; Khoi, H.H. 1999. Effect of daily and weekly micronutrient supplementation on micronutrient deficiencies and growth in young Vietnamese children. American Journal of Clinical Nutrition, 69, 80–86.

Udomkesmalee, E.; Dhanamitta, S.; Sirisinha, S.; Charoenkiatkul, S.; Tuntipopipat, S.; Banjong, O.; Rojroongwasinkul, N.; Kramer, T.R.; Smith, J.C., Jr. 1992. Effect of vitamin A and zinc supplementation on the nutriture of children in Northeast Thailand. American Journal of Clinical Nutrition, 56, 50–57.

UN (United Nations). 1994. WISTAT — women's indicators and statistics database of the United Nations [CD-ROM, version 3]. United Nations Publications, New York, NY, USA.

UNICEF (United Nations Children's Fund). 1999. State of the world's children. United Nations, New York, NY, USA. 131 pp.

USDA (US Department of Agriculture, Agricultural Research Service). 1999. USDA nutrient database for standard reference (release 13) [database online]. Nutrient Data Laboratory, USDA, Beltsville, MD, USA. Available from http://www.nal.usda.gov/fnic/foodcomp/. Cited September 1999.

Valberg, L.S.; Flanagan, P.R.; Chamberlain, M.J. 1984. Effects of iron, tin, and copper on zinc absorption in humans. American Journal of Clinical Nutrition, 40, 536–541.

Walravens, P.A.; Chakar, A.; Mokni, R.; Denise, J.; Lemonnier, D. 1992. Zinc supplements in breastfed infants. Lancet, 340, 683–685.

Walravens, P.A.; Hambidge, K.M. 1976. Growth of infants fed a zinc supplemented formula. American Journal of Clinical Nutrition, 29, 1114–1121.

Walravens, P.A.; Hambidge, K.M.; Kopfer, D.M. 1989. Zinc supplementation in infants with a nutritional pattern of failure to thrive: a double-blind, controlled study. Pediatrics, 83, 532–538.

Walravens, P.A.; Krebs, N.F.; Hambidge, K.M. 1983. Linear growth of low income preschool children receiving a zinc supplement. American Journal of Clinical Nutrition, 38, 195–201.

WHO (World Health Organization). 1996. Trace elements in human nutrition and health. WHO, Geneva, Switzerland. 343 pp.

Yadrick, M.K.; Kenney, M.A.; Winterfeldt, E.A. 1989. Iron, copper, and zinc status: response to supplementation with zinc or zinc and iron in adult females. American Journal of Clinical Nutrition, 49, 145–150.

Zinc Investigators' Collaborative Group (Bhutta, Z.A.; Black, R.E.; Brown, K.H.; Meeks Gardner, J.; Gore, S.; Hidayat, A.; Khatun, F.; Martorell, R.; Ninh, N.X.; Penny, M.E.; Rosado, J.L.; Roy, S.K.; Ruel, M.; Sazawal, S.; Shankar, A.). 2000. Therapeutic effects of oral zinc in acute and persistent diarrhea in children in developing countries: pooled analysis of randomized controlled trials. American Journal of Clinical Nutrition. In press.

The Editors

Kenneth H. Brown, MD, is director of the Program in International Nutrition and professor in the Department of Nutrition at the University of California, Davis. He is also adjunct professor in the Department of International Health at The Johns Hopkins University in Baltimore. Dr Brown was previously director of research at the Instituto de Investigación Nutricional in Lima, Peru, and director of the Division of Human Nutrition in the Department of International Health at The Johns Hopkins University. From 1992 to 1994, Dr Brown was president of the Society for International Nutrition Research. Among his numerous appointments, Dr Brown currently sits on the Editorial Board of the *Journal of Health, Population and Nutrition* and chairs the Steering Committee of the International Zinc Nutrition Consultative Group. Dr Brown is a past recipient of the Kellogg International Nutrition Research Prize (presented by the Society for International Nutrition Research) and the E.V. McCollum Award (presented by the American Society for Clinical Nutrition).

Sara E. Wuehler is a registered dietician and doctoral student in the Program in International Nutrition at the University of California, Davis. From 1991 to 1994, she worked as a volunteer for the United States Peace Corps in Gabon. From 1994 to 1997, Ms Wuehler was dietician and clinical director of Women, Infants and Children (WIC) in Provo, Utah.

The Publisher

The Micronutrient Initiative (MI) was established in 1992 as an international secretariat by its principal sponsors: the Canadian International Development Agency (CIDA), Canada's International Development Research Centre (IDRC), the United Nations Children's Fund (UNICEF), the United Nations Development Programme (UNDP), and the World Bank. MI's mission is to facilitate the following goals: virtual elimination of iodine deficiency disorders, virtual elimination of vitamin A deficiency and its consequences, including blindness, and reduction of iron deficiency anemia in women by one-third of the 1990 levels. It does so by supporting effective and sustainable programs in five areas considered critical to national and global efforts to eliminate micronutrient malnutrition: advocacy and alliance building, development and application of technologies, regional and national initiatives, capacity building, and resolution of key operational issues.

This document is published by MI in association with IDRC, a public corporation created by the Parliament of Canada in 1970 to help developing countries use science and knowledge to find practical, long-term solutions to the social, economic, and environmental problems they face. Support is directed toward developing an indigenous research capacity to sustain the policies and technologies that developing countries need to build healthier, more equitable, and more prosperous societies. IDRC publications are sold through its head office in Ottawa, Canada, as well as by IDRC's agents and distributors around the world. The full catalogue is available at http://www.idrc.ca/booktique/.